# PREFACE

Research, development, and demonstration of oxygenated and alternative fuels in transportation applications are vital for both environmental and energy strategies as we progress into the 21st century. Pollutant emissions from internal combustion engines have been lowered through advances in engine controls and aftertreatment technologies. Alternative fuels not only provide another path to reduced emissions with the added benefit of decreasing our dependence on foreign oil, but may enhance the performance of after treatment systems. Oxygenated and alternative fuels provide an important direction for achieving further emission reductions while still providing low-cost and fuel-efficient designs. The papers in this publication include a diverse group of subjects such as enhancing oxygenated additives like 1,3-dioxolane and di-tertiary-butyl peroxide (DTBP) and alternative fuels such as natural gas, propane, and a mixture of a biomass-based gas blended with natural gas. These diverse papers looked at the impact of these fuel technologies on engine properties, exhaust emissions, and performance; and cover a wide range of work from researchers throughout the world in both spark-ignited and compression-ignition engines.

This SAE Special Publication, *Research in Alternative Fuel Development* (SP-1716), includes papers from the Alternate Fuels sessions at the 2002 Spring Fuels & Lubricants Meeting & Exhibition. Technologies include a diverse group of subjects in oxygenates such as dimethyl ether (DME), 1,3-dioxolane, di-tertiary-butyl peroxide (DTBP), and dimethoxymethane and in alternative fuels such as natural gas, propane, and a mixture of a biomass-based gas blended with natural gas. Several papers looked at the impact of these fuel technologies on safety, handling, and engine properties such as spray characteristics, exhaust emissions, and performance.

**Michael D. Kass**
Oak Ridge National Laboratory


**E. Robert Fanick**
Southwest Research Institute


Session Organizers

# Research in Alternative Fuel Development

## SP-1716

**MOBILITY** *DATABASE*

*ers, standards, and selected
stracted and indexed in the
lity Database*

Published by:
motive Engineers, Inc.
Commonwealth Drive
ndale, PA 15096-0001
USA
Phone: (724) 776-4841
Fax: (724) 776-5760
May 2002

ISBN 0-7680-1008-X
SAE/SP-02/1716
Library of Congress Catalog Card Number: N98-42939
Copyright © 2002 Society of Automotive Engineers, Inc.

# TABLE OF CONTENTS

2001-01-3680

# Performance and Emissions of a DI Diesel Engine Operated with LPG and Ignition Improving Additives

**M. Alam and S. Goto**
National Institute of Advanced Industrial Science and Technology

**K. Sugiyama, M. Kajiwara and M. Mori**
Iwatani International Corp.

**M. Konno, M. Motohashi and K. Oyama**
Ibaraki Univ.

## ABSTRACT

This research investigated the performance and emissions of a direct injection (DI) Diesel engine operated on 100% butane liquid petroleum gas (LPG). The LPG has a low cetane number, therefore di-tertiary-butyl peroxide (DTBP) and aliphatic hydrocarbon (AHC) were added to the LPG (100% butane) to enhance cetane number. With the cetane improver, stable Diesel engine operation over a wide range of the engine loads was possible. By changing the concentration of DTBP and AHC several different LPG blended fuels were obtained. In-cylinder visualization was also used in this research to check the combustion behavior. LPG and only AHC blended fuel showed $NO_X$ emission increased compared to Diesel fuel operation. Experimental result showed that the thermal efficiency of LPG powered Diesel engine was comparable to Diesel fuel operation. Exhaust emissions measurements showed that $NO_X$ and smoke could be considerably reduced with the blend of LPG, DTBP and AHC.

## INTRODUCTION

For future energy supplies, it will be necessary utilize remote natural gas sources and transport the gas by pipeline over longer distances than currently economically feasible. DME (Dimethyl Ether), methanol and LPG are especially interesting when the energy is needed or desired as a transportation fuel, and when it has to be transported over increasingly long distances.

Diesel engine is considered one of the most environmental friendly engines due to the longer travel distance with the same amount of fuel, with lower carbon dioxide ($CO_2$) emissions (one of the major greenhouse gases associated with the global warming). As far as emission is concerned, one of the promising ways to reduce emission of diesel engine is the use of alternative fuel such as methanol, DME, GTL (Gas to Liquid), CNG (Compressed Natural Gas) and LPG [1 to 5].

Among the alternative gaseous fuels on the market, LPG and CNG are most widely used as a fuel for SI engines because of their low cetane number. $NO_X$ emission level of an LPG SI engine is high at stoichiometric operation. Also the high exhaust gas temperature at stoichiometric operation causes durability problem and lower thermal efficiency for an LPG powered SI engine.

When Lean burn LPG SI engine is considered, then the lower exhaust gas temperature improves the durability, $NO_X$ emissions, and thermal efficiency considerably [6, 7]. However, lean burn LPG SI engine has some drawbacks. Especially, burning velocity at lean LPG fuel air mixtures is significantly lower than at stoichiometric operation. This might increase overall combustion duration, resulting in an increase heat transfer loss through the cylinder wall and decrease overall thermal efficiency of the engine. Moreover, lower burning velocity leads to engine misfire, which in-turn causes cycle-to-cycle variation of the indicated work and high levels of unburned hydrocarbon emissions [8 to 10].

The price of LPG fuel is lower than the Diesel fuel. LPG has been primarily a by-product of crude oil refining process, however recently more LPG is being produced as a by-product of natural gas. Therefore, more stable supply of LPG is possible, as is the case with natural gas, compared to Diesel. In Japan, more LPG SI vehicles have been coming into operation in recent years. In Japan, there are more than two thousand refueling stations nationwide, which is a big advantage in available infrastructure when this LPG is used more widely [11].

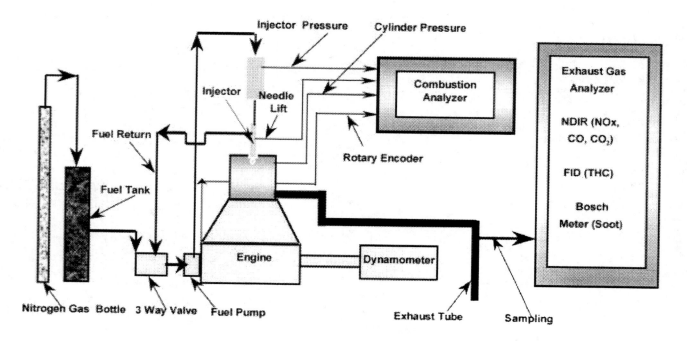

Figure 1. Schematic diagram of the experimental system.

LPG has been primarily used as fuel for SI engines rather than compression ignition (CI) engines, due to the low cetane number of LPG. A previous report [12] showed that enhancing cetane number of LPG is a suitable method for operating a DI Diesel over a wide range of engine loads. Consequently, high thermal efficiency can be achieved with CI engine compared to the operation of SI with LPG. Moreover, due to the low molecular weight and high vapor pressure of LPG, homogeneous fuel/air mixtures followed by premixed gaseous combustion can easily be attained. Thus, operation of a Diesel engine that emits significantly less soot and $CO_2$ on a calorific basis as compared to a conventional Diesel fueled engine is possible using LPG fuel.

Therefore, in this research of LPG powered DI Diesel engine, experiments were conducted to enhance the cetane number by blending cetane improvers such as DTBP and AHC to the LPG, to improve the auto-ignition characteristics and achieve stable engine operation for a wide range of engine loads. The effects of additive concentrations on engine performances and emissions were compared to a conventional Diesel fueled engine.

## EXPERIMENTAL

ENGINE AND APPARATUS - Experiments were conducted with a single cylinder, naturally aspirated, four stroke, water cooled, direct injection research Diesel engine model 6D22 (T6) manufactured by Mitsubishi Motors, which is actually used in ten-ton class trucks. A jerk type injection pump was used with a plunger diameter of 12mm, due to lower calorific value (J/ml) of LPG compared to Diesel fuel. The engine Specifications are shown in Table 1.

Table 1. Engine specifications

| Bore × Stroke | 130 × 140 |
|---|---|
| Displacement | 1858 cm$^3$ |
| Valves | OHV, 2 Valves |
| Type | 4 stroke, Diesel |
| Fuel Delivery | Direct Injection |
| Compression Ratio | 17.0 |
| Piston Cavity | Re-entrant Toroidal |
| Plunger Diameter | 12 mm |
| Injection Pressure | 24.5 MPa (Diesel Fuel) 9.8 MPa (LPG Blend) |

Figure 1 presents a schematic diagram of the experimental system. The engine was sufficiently instrumented with several measuring devices for implementing the objective of the study, including installation of the needle lift sensors, pressure transducers and interfacing the engines with emission measurement devices and combustion analyzer. LPG has a high vapor pressure in the laboratory environment, therefore the fuel system was pressurized at 1.50MPa by nitrogen gas that was supplied from a gas cylinder in order to prevent leakage, cavitations and vapor lock of the fuel system. During operation with LPG blend, injector opening pressure was set at 9.8Mpa. For Diesel fuel, injector opening pressure was 24.5MPa, which was recommended by the engine manufacturer.

The cost of butane is lower than that of propane. The LPG fuel used in this experiment was 100% butane, which was consisted of 70% *n*-butane and 30% *i*-butane. Table 2 shows the characteristics of LPG used in this work. In the U.S., vehicle LPG almost always refers to "Propane HD - 5" LPG. The specification for HD - 5 is that it contains at least 90% propane. Therefore, LPG used in this research is different from the U. S. "Propane HD - 5" fuel.

Table 2. Properties of LPG

|  | *i*-butane | *n*-butane |
|---|---|---|
| Molecular Formula | $C_4H_{10}$ | $C_4H_{10}$ |
| Boiling Point ($^o$C) | -11.7 | -0.5 |
| Research Octane number | 102 | 94 |
| Stoichiometric A/F (kg/kg) | 15.49 | 15.49 |
| Auto ignition Temperature ($^o$C) | 544 | 441 |
| Lower Heating Value (MJ/kg) | 45.55 | 45.70 |

In these experiments, several cetane enhanced LPG blended fuels (composition of blends will be discussed later) were used to check the combustion performances and emission characteristics in a DI Diesel engine and compare these data with Diesel fuel operation. Experimental conditions are shown in Table 3. The cylinder pressure was measured by using a piezoelectric transducer for 350 cycles with 0.5 crank angle degree resolution. Measured data were recorded by a combustion analyzer model CB566, which was also computed the cycle variation via the variation of indicated mean effective pressure. An exhaust gas analyzer model BEX-5100D was used to measure $NO_X$, THC (Total Hydrocarbon), CO, $CO_2$ and $O_2$ emissions. A Bosch smoke meter was also used to measure the smoke emission. The fuel flow rate was calculated from the measured $CO_2$ concentration using the carbon balance method [13].

Table 3. Experimental conditions

| Engine Speed | 1500 RPM |
|---|---|
| Injection Timing | 15 deg BTDC |
| Torque | 0.03 ~ 0.6 MPa |

CETANE ENHANCED LPG BLENDED FUELS - DTBP (Di-tertiary-butyl peroxide) and AHC (aliphatic hydrocarbons) were blended to LPG for enhancing cetane number, since the LPG is a low cetane fuel. Details of the LPG blended fuels and their static injection timing are shown in Table 4. By changing the concentration of DTBP and AHC several different LPG blended fuels were obtained. In the following section, blend of LPG + 15% DTBP, LPG + AHC and LPG + AHC + 1wt% DTBP will be denoted as LPG A, LPG B and LPG C respectively (Table 4).

For LPG fuel, there is a time delay between the static (set for injection) and dynamic (actual needle movement) injection of fuel. The compressibility of LPG is larger than Diesel fuel, which makes delay in the

dynamic injection timing of LPG. Details of the interaction between static and dynamic injection timing can be found in the earlier study [12] therefore, no further discussion is added here. In this experiment, static injection timing for each LPG fuels were set in a different position (Table 4) to perform dynamic injection with the LPG fuels are very close to the Diesel fuel. The static injection timing of Diesel fuel was set at 18 degrees before top dead center (BTDC).

Table 4 LPG blends and their static injection timing

| Cetane Enhanced LPG Blends | Fuel Blend Name | Static Injection Timing, deg BTDC |
|---|---|---|
| LPG + 15wt% DTBP | LPG A | - 21 |
| LPG + AHC* | LPG B | - 20 |
| LPG + AHC* + 1wt% DTBP | LPG C | - 27 |
| *Particular name and concentration of AHC and DTBP added to the LPG were not shown here due to the restriction from the sponsor | | |

**Cetane Number = - 47.48T + 124.8**

Where T is the ignition delay in ms.

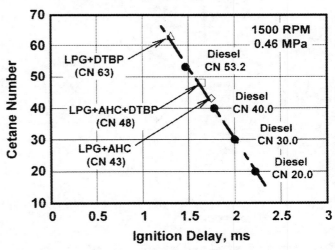

Figure 2. Relationship between ignition delay and cetane number.

Figure 2 presents cetane number with the variation of ignition delay. Black circles indicate four types of Diesel fuels with known cetane numbers of, 20, 30, 40 and 53.2. Experiments were conducted to know the ignition delay of the Diesel fuels with known cetane number, and the above relation for cetane number and ignition delay was determined:

Therefore, by knowing the ignition delay, relative cetane number of the LPG blended fuels can be find out from the above relation. The cetane number of the LPG A, LPG B and LPG C fuels used in this research were 63, 43 and 48 respectively.

LUBRICATING ADDITIVE – A small amount of fuel additive (long-chain alkyl ester) was added to the LPG blends to reduce the wear on the needle of the injector nozzle when LPG was used, due to the lower lubricity of LPG compared to Diesel fuel. The fuel additive did not appear to alter the combustion performance and emission characteristics of the engine.

## RESULTS

The results obtained in this study concerning engine performance and emission characteristics are presented in Figs. 3 to 13, where in the most cases, solid line indicates Diesel and dotted lines represent LPG blended fuels.

COMBUSTION PERFORMANCE - Figs. 3 and 4 represent the needle lift and cylinder pressure characteristics respectively, of all the fuels used in this experiment. The operating load condition is 0.46MPa, fixed engine speed at 1500rpm.

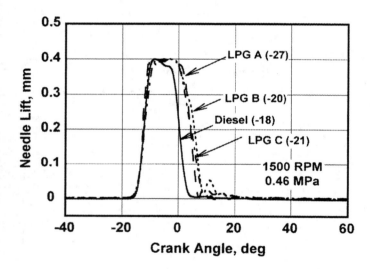

Figure 3. The characteristic of needle lift with Diesel and LPG blended fuels.

The main advantage of the LPG is lower cost compared to Diesel fuel. However, a major disadvantage of the LPG is low cetane rating when operation in a compression ignition (CI) engine is considered to utilize the higher thermal efficiency of the CI engine. By blending DTBP, cetane rating can be enhanced. However, cetane enhanced LPG fuel results in increased cost, and more than 5wt% DTBP is required for stable engine operation over a wide range of engine loads [12]. Therefore, in this research AHC was added to the LPG to reduce cetane enhanced cost so that a cost effective LPG blend fuel can be introduced for CI engine operation.

As explained earlier, static injection timing for Diesel and LPG blends were not set at the same crank angle since the compressibility of LPG is higher than the Diesel fuel. The more LPG in the blend, the more static injection timing should be advanced, if the dynamic start of injection timing of LPG blends is to be the same as Diesel fuel. Figure 3 shows dynamic needle lift start with the Diesel and LPG blends are almost at the same crank angle (CA). Another difference between the LPG blends and Diesel fuel is injection duration. Injection duration with the LPG blended fuels is higher than that of Diesel fuel, which means more fuel is required to deliver same power with the LPG blends compared to Diesel fuel.

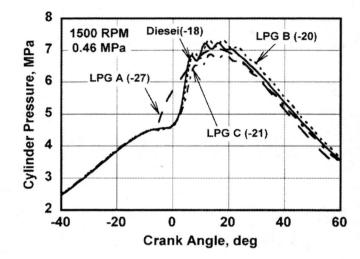

Figure 4. The characteristic of cylinder pressure for Diesel and LPG blends.

Figure 4 represents cylinder pressure with the variation of crank angle for Diesel and LPG blended fuels. High cetane rating (Fig. 2) of LPG A showed early start of combustion compared to other fuels. Rest of the fuel showed combustion start just before the TDC. Maximum peak pressure was observed with LPG B blend and LPG C showed lowest peak cylinder pressure.

IGNITION DELAY – A combination of physical and chemical delay comprises ignition delay of the fuel. With increase in engine speed, both physical and chemical delay can be reduced since turbulence and compression temperature increase. High compression ratio as well as high cetane rating reduces chemical delay, which in turn reduces ignition delay.

A comparative study of the ignition delay for Diesel and LPG blends is shown in Fig. 5. The figure also shows the effect of variation of engine load on ignition delay. In general, ignition delay decreases almost linearly with increase in load, due to the increase of residual gas temperature and cylinder wall temperature. Approximate cetane ratings of the fuels used in this research were in the order of: LPG A > Diesel > LPG C > LPG B (Fig. 2). Ignition delay with the variation of engine load also correlated to cetane number; that is, the higher the cetane rating, the lower the ignition delay of the fuels. LPG A showed lowest ignition delay, and LPG B showed highest. The cetane rating of the two fuels were 63 and 43 respectively (Fig. 2).

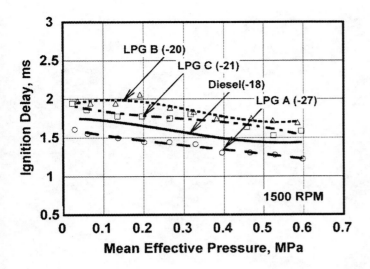

Figure 5. Comparison of Ignition delay with the variation of engine load.

COMBUSTION VISUALIZATION – Flame images were taken by a high-speed video camera (Kodak Ekta Pro HS 4540) with an image intensifier (IMCO PLS-3) at 4500 frames per second. Figure 6 presents the actual view of the test engine used for combustion visualization. The same test engine was modified for in-cylinder visualization, where pictures were taken by using an extended bottom view piston having a quartz window. The time interval between frames was about 0.22ms at an engine speed of 1200rpm. The engine was operated with an external load of 0.07MPa. Table 5 presents operating condition of the experiment.

Figure 6. A single cylinder test engine for in-cylinder combustion visualization

The thermal efficiency (Fig. 8) and exhaust emissions measurement with LPG C showed better results than the Diesel. Therefore, a study was undertaken to determine the in-cylinder combustion differences between Diesel and LPG C. Experimental conditions used (Table 5) for in-cylinder combustion visualization were a bit different from the parameters used during performance test of the engine. Figure 7 represents consecutive images of the in-cylinder combustion of Diesel and LPG C. The light intensity of LPG C combustion was higher than of Diesel fuel. Moreover, combustion was completed at the center of the cylinder with Diesel, whereas with LPG C, it was completed at the periphery of the cylinder.

Table 5. Experimental conditions

| Engine Speed | 1200 RPM |
|---|---|
| BMEP | 0.07 MPa |
| Injection Timing | 15 deg BTDC |
| Injection Pressure | 16.3 MPa |
| Frame Speed | 4500 frames/sec |
| I.I. Gain | 45 % |
| Exposure Time | 0.1 μs |
| Delay | 0.04 μs |

PERFORMANCE AND EMISSIONS – Three types of cetane enhanced LPG blends were used in this research to compare the engine performance and exhaust emissions characteristics with Diesel fuel. Thermal efficiency, exhaust gas temperature, $NO_X$, THC (total hydrocarbon) and CO emissions will be discussed in this section.

By varying the engine load, brake thermal efficiency was obtained for Diesel and LPG blends, as shown in Fig. 8. LPG C showed relatively higher thermal efficiency at high load compared to Diesel fuel. The lowest thermal efficiency was obtained with the high cetane rating LPG blended fuels such as LPG A. As shown in the figure, the brake thermal efficiency remains below 40% due to the high mechanical loses of the test engines.

Exhaust gas temperature with the variation of external engine load is shown in Fig. 9. Exhaust gas temperature with Diesel and LPG B fuel was almost identical. LPG C blend showed highest exhaust gas temperature among the fuels. It should be noted that decreasing cetane rating should increase the ignition lag duration resulting in the late combustion, which in turn results in relatively higher exhaust gas temperature. However, different results were obtained, as shown in the above plot. Exhaust emissions of the automotive engine are the major problem we are facing today. However, emission reduction with Otto cycle engine is much easier compared to diesel cycle engine when catalytic reduction is considered. Soot formation with Otto cycle engine is very low, or negligible, compared to Diesel engine. $NO_X$ and particulate emissions are the severest problem when Diesel engine is considered. Engine exhaust emissions $NO_X$, THC, CO and smoke were measured for all the fuel used in this study. Figs. 10 to

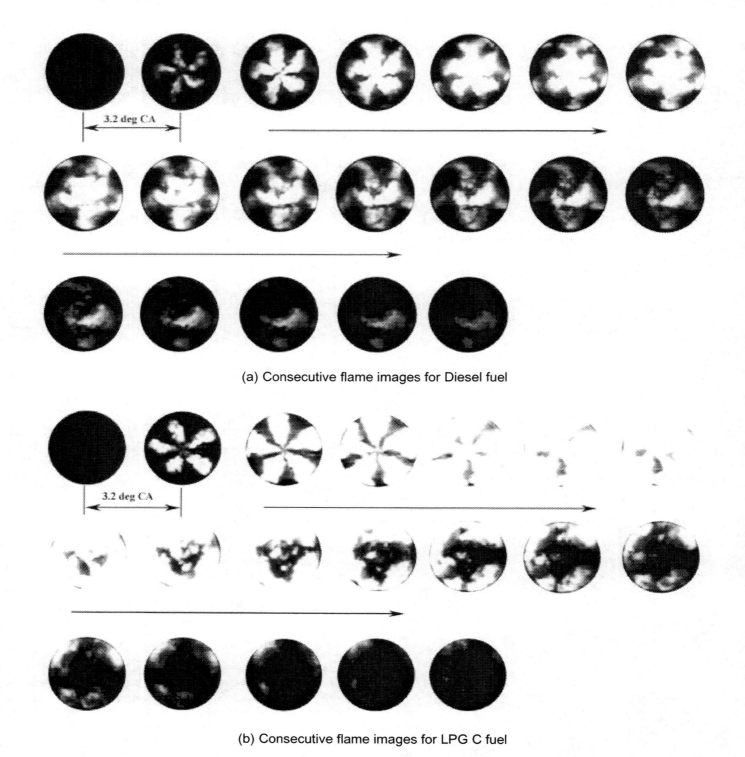

(a) Consecutive flame images for Diesel fuel

(b) Consecutive flame images for LPG C fuel

Figure 7. Comparison of in-cylinder combustion visualization with (a) Diesel fuel and (b) LPG C

13 present the emission characteristics which are plotted as a function of mean effective pressure (MEP).

$NO_x$ emissions with Diesel and cetane enhanced LPG blended fuels are shown in Fig. 10. From this plot, $NO_x$ emissions generally increased with increase in brake mean effective pressure. The lowest $NO_x$ emission was observed with LPG C compared to the other fuels. This is possibly due to the lower cylinder pressure during

combustion. The LPG B showed higher $NO_x$ emissions than that of Diesel, due to higher cylinder pressure. The high $NO_x$ emission with LPG A compared to Diesel may be caused by the early start of combustion (Fig. 4) and more after compression of combustion products, which results in a higher temperature.

THC emissions are the consequence of incomplete combustion of hydrocarbons in the fuel. The inability to complete the combustion of the fuel lowers the combustion efficiency. The primary aim of combustion engineering is to maximize the combustion efficiency by minimizing the emission of hydrocarbons. The most common cause for incomplete burning of fuel is insufficient mixing between the fuel, air, and combustion products. The relative concentrations of hydrocarbon emissions are greatly influenced by the composition of the fuel. Figure 11 presents the THC data with the variation of engine load.

THC emissions with Diesel operation were about the amount expected from a typical diesel engine. LPG A showed very small amount of THC emission for a wide range of engine loads. LPG B showed higher THC emissions compared to Diesel. In the cylinder LPG blends are expected to disperse locally to form lean fuel spray pattern. This may be a factor in consuming the maximum amount of fuel before being exhausted from the cylinder.

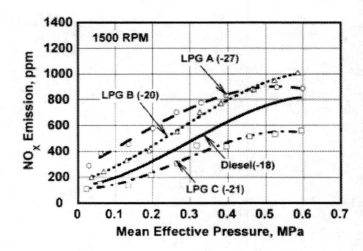

Figure 10. Comparison of engine out NO$_X$ emission with the variation of engine load.

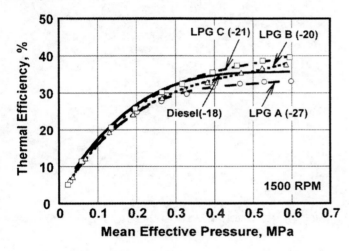

Figure 8. Brake thermal efficiency with the variation of engine load.

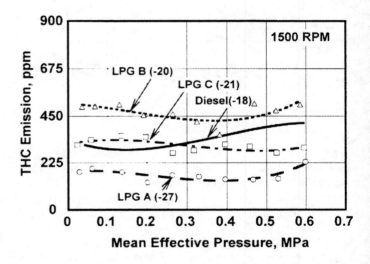

Figure 11. Comparison of engine out THC emission with the variation of engine load.

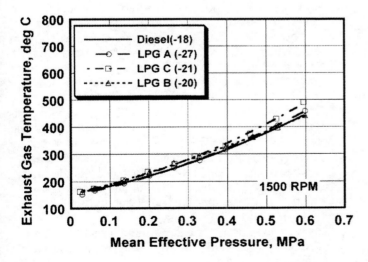

Figure 9. Exhaust gas temperature with the Diesel and cetane enhanced LPG blends.

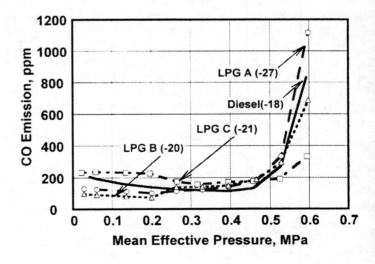

Figure 12. Comparison of engine out CO emission with the variation of engine load.

Carbon monoxide (CO) is formed as an intermediate species in the oxidation of carbon containing fuels. CO formation is one of the principal reaction paths in the hydrocarbon combustion mechanism. CO emissions, measured with the variation of engine load, are shown in Fig.10. CO emissions increased at high MEP with all the fuels due to the incomplete oxidation of fuels.

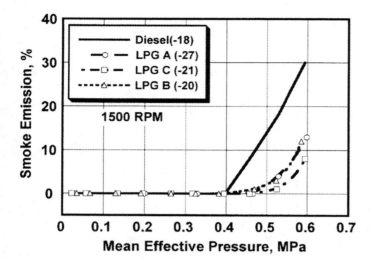

Figure 13. Comparison of engine out smoke emission with the variation of engine load.

The emission of smoke with diesel and cetane enhanced LPG blends is shown in Figure 13. Smoke formation was not observed with the fuels for brake mean effective pressure below 0.4MPa. However, high smoke emission was observed after 0.40MPa with Diesel compared to LPG blend operation. Therefore, it can be concluded that the smoke emissions of an LPG Diesel engine are very low compared to Diesel fuel operation. The absence of rich fuel pockets, or any exposure of such mixture to high temperature combustion products, is a precondition for low smoke emission from a conventional Diesel engine. The formation of a fuel-rich mixture may not be likely in LPG spray plumes since the fuel is highly volatile, becoming a gaseous jet and attaining high specific volumes as soon after leaving the injector. The low smoke formation with LPG blends in a Diesel engine may suggest that the spray formation with cetane enhanced LPG blended fuels are sufficiently lean to resist smoke formation.

ACTUAL LPG DIESEL TRUCK – Experimental results showed that operation with the LPG C reduced emissions, especially $NO_X$ and smoke. Moreover, thermal efficiency was comparable to Diesel fuel. Therefore, an actual Diesel truck with in-line plunger pump has been modified for operating with the cetane enhanced LPG blend. Another two trucks of same specifications, but built with rotary distributor pumps have been developed. Modifications include fuel tank for LPG, $N_2$ gas cylinder for pressurization of the LPG, oxidation catalyst for unburned hydrocarbons and de-$NO_X$ catalyst for $NO_X$ reduction.

Table 6. Specifications of the LPG Diesel truck

| Bore × Stroke | 108 × 115mm |
|---|---|
| Compression Ratio | 18 |
| Number of Cylinder | 4 |
| Plunger Diameter | 10mm |
| Load Capacity | 2000kg |
| Engine Output | 130PS @ 3200RPM |

Specifications of the truck are shown in Table 6. Figure 14 shows an actual direct injection Diesel truck operated with LPG C. It has been operating for about 4000km over the last year without any major failure.

Figure 14. An external view of the prototype LPG Diesel truck developed in this work.

## CONCLUSION

To investigate low emissions producing alternative fuels, experiments were conducted in a typical direct-injection Diesel engine by blending cetane improver such as, DTBP and AHC with LPG to improve the auto-ignition characteristics and achieve stable engine operation for a wide range of engine loads. Some of the findings from the present study are summarized in the following:

1. Enhancing cetane rating of LPG by blending di-tertiary-butyl peroxide (DTBP) and aliphatic hydrocarbons (AHC) is a plausible method to operate a direct injection Diesel engine. By adding AHC and 1% DTBP, a cost effective LPG blend can be introduced for operating a conventional DI Diesel engine for a wide range of engine loads.

2. Among the blends of LPG used in this research, LPG C showed brake thermal efficiency comparable to Diesel fuel operation.

3. Exhaust emissions measurement showed that $NO_X$ and smoke emissions could be reduced with LPG C.

4. The light intensity of LPG C combustion was higher than of Diesel fuel. Moreover, combustion was completed at the center of the cylinder with Diesel, whereas LPG combustion was completed at the periphery of the cylinder.

## ACKNOWLEDGMENTS

The authors wish to express their appreciation to Mr. Makoto Sagara of Iwatani International Corporation who assisted the experiment.

## REFERENCES

1. Eberhard, G. A., Ansari, M., and Hoekman, S. K., "Emissions and Fuel Economy Tests of a Methanol Bus with a 1988 DDC Engine", SAE Paper 900342, 1990.
2. Christensen, R., Sorenson, S. C., Jensen, M. G., and Hansen, K. F., "Engine Operation on Dimethyl Ether in a Naturally Aspirated, DI Diesel Engine", SAE Paper 971665, 1997.
3. Suga, T., Kitajima, S., and Fujii, I., "Pre-Ignition Phenomena of Methanol Fuel (M85) by the Post-Ignition Technique", SAE Paper 892061, 1989.
4. Kajitani, S., Chen, C. L., Oguma, M., Alam, M., and Rhee, K. T., "Direct Injection Diesel Engine Operated with Propane – DME Blended Fuel", SAE Paper 982536, 1998.
5. Stavinoha, L. L., Alfaro, E. S., Dobbs, H. H., Villahermosa, L. A., and Heywood, J. B., "Alternative Fuels: Gas to Liquids as Potential 21st Century Truck Fuels", SAE Paper 2000-01-3422, 2000
6. Lee, D., Shakal, J., Goto, S., Ishikawa, H., Ueno, H., and Harayama, N., "Observation of Flame Propagation in an LPG Lean Burn SI Engine", SAE Paper 1999-01-0570, 1999.
7. Goto, S., Lee, D., Shakal, J., Harayama, N., Honjo, F., H., and Ueno, H., "Performance and Emissions of an LPG Lean Burn Engine for Heavy Duty Vehicles", SAE Paper 1999-01-1513, 1999.
8. Witze, P. O., and Vilchis, F. R., "Stroboscopic Laser Shadow Graph Study of the Effect of Swirl on Homogeneous Combustion In a Spark-Ignition Engines", SAE Paper 830335, 1983.
9. Ando, H., "Combustion Control Technologies for Gasoline Engine", ImechE Seminar Publication 1996-20, 1996.
10. Belaire, R. C., Davis, G. C., Kent, J. C., and Tabaczynski, R. J., "Combustion Chamber Effects on Burn Rates in a High Swirl Spark Ignition Engines", SAE Paper 830335, 1983.
11. Goto, S., Shakal, J., and Daigo, H., "LPG Engine Research and Development in Japan", International Congress on Transportation Electronics (Convergence '96), 1996.
12. Goto, S., Lee, D., Wakao, Y., Honma, H., Mori, M., Akasaka, Y., Hashimoto, K., Motohashi, M., and Konno, M., "Development of an LPG DI Diesel Engine Using Cetane Number Enhancing Additives", SAE Paper 1999-01-3602, 1999.
13. Heywood, J. B., Internal Combustion Engine Fundamentals, McGrow-Hill, 1998.

2001-01-3682

# Effect of Reduced Boost Air Temperature on Knock Limited Brake Mean Effective Pressure (BMEP)

**Chad Stovell and James Chiu**
Southwest Research Institute

**Ralph Hise and Paul Swenson**
Advanced Technologies Management Inc.

## ABSTRACT

The effect of low temperature intake air on the knock limited brake mean effective pressure (BMEP) in a spark ignited natural gas engine is described in this paper. This work was conducted to demonstrate the feasibility of using the vaporization of liquefied natural gas (LNG) to reduce the intake air temperature of engines operating on LNG fuel. The effect on steady-state emissions and transient response are also reported. Three different intake air temperatures were tested and evaluated as to their impact upon engine performance and gaseous emissions output. The results of these tests are as follows. The reduced intake air temperature allowed for a 30.7% (501 kPa) increase in the knock-limited BMEP (comparing the 10°C (50°F) intake air results with the 54.4°C (130°F) results). Exhaust emissions were recorded at constant BMEP for varying intake air temperatures. The raw $BSNO_x$ emissions were reduced by 0.6% (0.1 g/kW-hr, not a significant change) due to the chilled air, but the corrected $BSNO_x$ emissions were reduced by 17.8% (3.4 g/kW-hr). The BSHC emissions were reduced by 39.9% (0.5 g/kW-hr). The BSCO emissions, however, were increased by 7.5% (0.2 g/kW-hr). Transient engine response revealed that the use of reduced temperature intake air provided a 15.3% (1.43 second) improvement in the time required to obtain an engine speed of 1600 rpm. Similarly, the chilled intake air reduced the time required to reach a set manifold air pressure of 212.4 kPa (30.8 psia) by 11.5% (1.21 seconds).

## BACKGROUND

One of the limitations of heavy-duty natural gas fueled engines is knock. A lowered intake air temperature could allow for higher power density or higher Brake Mean Effective Pressure (BMEP) for heavy-duty natural gas fueled engines [1]. One of the effects that govern the phenomenon of knock is the temperature history of the air and fuel mixture.

One method to achieve low intake air temperatures is to utilize the vaporization of Liquefied Natural Gas (LNG) on a natural gas fueled engine. The lowest possible intake temperature on an engine with an air-to-air intercooler is ambient. Fueling systems used on LNG vehicles have a heat exchanger that typically uses engine coolant to vaporize the fuel before being delivered to the engine. This heat transfer uses energy from the engine's coolant to bring the LNG from a liquid state to a vapor state. Instead of using engine coolant to vaporize the LNG, the hot compressed air from the turbocharger can be used in this "LNG air cooler", which can lower the intake air temperature below ambient.

The goal of this project was to conduct a series of experiments to simulate a system that would use the vaporization of LNG to cool the intake air to the engine. This simulated system involved a chilled air intake system to observe the effect on knock limited BMEP, gaseous emissions, and transient performance. An investigation was conducted to find the difference in the BMEP at the onset of knock with 10°C (50°F), 26.6°C (80°F) and 54.4°C (130°F) intake air temperatures. Also, a simple transient test, a ramp test from idle to full load, was performed at 10°C (50°F) and 54.4°C (130°F) to provide a performance comparison. Southwest Research Institute was commissioned by Advanced Technologies Management Inc. (ATM) to carry out these experiments and report upon the findings. Funding for this project was provided by the U.S. Department of Energy Office of Heavy Vehicle Technologies.

## EXPERIMENTAL METHOD

A test was conducted to determine the effect of reduced intake air temperature on knock limited BMEP and on the accompanying gaseous emissions. This test was conducted on a Mack E7G natural gas fueled engine [2]. The intake system was equipped with an additional heat exchanger and a chiller, which simulated the cooling effect associated with the vaporization of the LNG. By controlling the temperature of the chilled water, the intake air's temperature was altered. During the testing process, three different temperatures were investigated. This section describes the equipment and procedure used to conduct this test.

## TEST SETUP

Test Engine - The engine used in this test was the Mack E7G. The Mack E7G is a production engine that is specifically designed to operate on natural gas. Engine specifications are listed below in Table 1.

Table 1. Engine Specifications

| Engine Model | E7G |
|---|---|
| Engine Displacement (L) | 12 |
| Bore and Stroke (mm) | 123.8 x 165.1 |
| Combustion | Lean Burn |
| Power Rating | 260 kW @ 1950 rpm |
| Torque Rating | 1695 N-m @ 1250 rpm |

A Woodward OH 1.2 engine control unit (ECU) controls the engine. This controller is interfaced with a standard PC to allow for user programmable capabilities. The OH 1.2 is a full authority engine controller which allows engine operating parameters such as ignition timing, equivalence ratio, and manifold air pressure, to be varied by the user while the engine is running. Ignition timing is controlled through a signal to a distributorless electronic ignition module. Equivalence ratio is controlled through a feedback system using a Universal Exhaust Gas Oxygen (UEGO) sensor. Manifold air pressure is controlled through wastegate actuation.

The engine is outfitted with a knock sensor and signal to noise enhancement filter (SNEF) module. The knock sensor is tuned for the frequency of knock for this engine. Signals from the sensor are then directed to the SNEF module that converts the signal from a vibration waveform to a step function. When the engine is not knocking, the SNEF module will output a 10V signal, and when the engine is knocking, the SNEF module detects knock and will output a 0V signal for 5 ms.

Engine Dynamometer - The engine is connected to an eddy current wet gap dynamometer. This dynamometer can govern either the steady-state speed or load of the engine. The dynamometer has a load cell that is capable of reporting the resultant torque. During the majority of the tests conducted in this project, steady state operation was the only mode that was used, however, a simple transient test was also performed on this engine. Although transient control of speed or load is not possible with the dynamometer and controller combination used for this project, transient power absorption is possible through the water drag on the wet

gap of the dynamometer. Further explanation of the use of the dynamometer for this simple transient test is detailed in the TEST PROCEDURE section.

Test Stand Setup - A single liquid-to-gas heat exchanger is a typical test cell component used with this engine to cool the intake air. This primary heat exchanger is meant to simulate the air-to-air intercooler found on the in-vehicle engine. The stock intercooler cools the high temperature boosted air by passing ambient air through the cooling fins. Through this process the boosted air is cooled down to a temperature ranging from approximately ambient to 54°C (130°F), depending upon vehicle speed, boost pressure, ambient temperature, etc. The test cell primary heat exchanger uses cooling tower water to provide the necessary cooling. This primary heat exchanger is equipped with an electronically controlled flow valve that regulates flow of water through the primary heat exchanger. The temperature set point for the air exiting the primary heat exchanger can be adjusted down to a temperature that is no lower than the temperature of the cooling tower water, ~29°C (85°F) depending on the ambient conditions.

Even though the primary heat exchanger can provide a substantial temperature drop in the intake air, it is still not sufficient for the temperatures required for this test. Due to this fact, a second heat exchanger was added to the air intake system. This second heat exchanger acts as the "LNG air cooler", and was supplied with chilled water from an external chilling unit. A Thermal Care ™ Accuchiller Model# AQ0A10 provided the continuous flow of low temperature liquid to the second heat exchanger. The Accuchiller unit directly controls the temperature of the water being supplied to the second heat exchanger. The supplied water's temperature can be set anywhere from −1°C (30°F) to 18.3°C (65°F).

Due to the low air temperatures exiting the second heat exchanger, and the high ambient temperatures and high relative ambient humidity, moisture will have a tendency to condense out and form liquid water in the intake system. To prevent this liquid water from entering the engine, a separator was placed in-between the second heat exchanger and the throttle. The separator was constructed from a 49.2 liter (13 gal.) cylindrical steel tank. The air enters the top of the separator and exits again through the top at a point on the opposite end of the separator. Any condensed liquid that enters the separator will fall to the bottom and remain there while the air exits out the top. A gate valve drain was installed on the bottom of the separator to allow for drainage of the trapped water. A simple test was run to determine the effectiveness of the separator and the results showed that ~73% of the condensed water was trapped by the separator. A simple diagram of the engine test stand setup is shown in Figure 1.

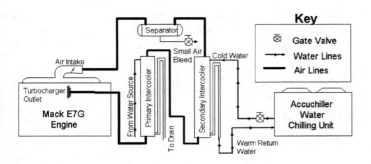

Figure 1. Engine Test Stand Setup

Test Fuel - The goal of this project was to observe the difference in knock limited BMEP between reduced intake air temperature and normal intake air temperature. LNG is typically 99% methane. This high percentage of methane results in a high octane rating, and therefore a high knock tolerance. With the chilled intake air, this knock tolerance might be even higher. To help induce knock, the fuel used for the knocking tests was specially blended to create a "worst case" LNG fuel. This type of fuel is characteristic of the LNG fuel used in Europe. The test fuel property targets and actual fuel properties used for this testing are listed in Table 2.

Table 2. Test Fuel

| Gas | Target (volume) | Actual (volume) |
|---|---|---|
| Methane | 85.0 ± 1.0% | 85.852% |
| Ethane/Ethylene | 9.0 ± 0.5% | 8.739% |
| Propane | 2.5 ± 0.5% | 2.294% |
| Nitrogen | 3.5 ± 0.5% | 3.102% |
| Other | | 0.013% |

Determination of Knock - As mentioned earlier, the engine has been outfitted with a knock sensor and SNEF module to identify knock. This instrument signifies when engine knock is occurring by outputting a 0V signal through the SNEF module. When the engine is not knocking the output from the SNEF module is a positive 10V signal. Signals from the SNEF module were directed to a digital I/O card on a stand-alone computer. A simple computer program was created to read in the incoming signal and integrate the value over time. When the signal remains high (10V), the integrated value reaches a value of 1.0. As the signal drops (to 0V) due to a knocking condition, the integrated value decays. The higher the occurrences of a "knock" signal, the more the integrated value will drop. A consistent knocking condition can now be identified by determining the corresponding integrated value. A trial run established that an integrated value (IV) of 0.87 corresponded to the onset of knock. This value was chosen as the determination between "knocking" and "not-knocking". By this method, all subjectivity was eliminated as to the "knocking" state of the engine. The knock sensor and

SNEF module used for this project are part of the production engine.

TEST PROCEDURE

Determination of "Knock" Point at 10°C (50°F) Intake Air Temperature - To test the concept of increased knock limit with reduced boost air temperature, the engine was run at 1300 rpm with certain engine parameters fixed (ignition timing, equivalence ratio, water out temperature, oil temperature, etc.). The highest manifold air pressure at this test condition before the onset of knock was determined for the lowest intake air temperature. This test point was determined by first increasing the manifold air pressure up to a maximum of 290 kPa (42 psia), then increasing ignition timing up to the maximum for best torque, with a maximum of 2169 N-m (1600 ft-lbs) of torque. The two limits on manifold air pressure and torque were then established as the safe limit for this testing. At a manifold air temperature of 10°C (50°F), a manifold air pressure of 290 kPa (42 psia), an equivalence ratio of 0.67, ignition timing of 24° BTDC and a torque of 2156 N-m (1590 ft-lb), the engine was not knocking. The next step was to increase equivalence ratio. The equivalence ratio was increased, with ignition timing at MBT and manifold air pressure set to produce 2169 N-m (1600 ft-lb) of torque. The engine began to knock at an equivalence ratio of 1.0 (stoichiometric). Therefore, a stoichiometric air/fuel ratio at 8° BTDC was selected as the test condition for all steady state testing.

Various Manifold Air Temperature Evaluations - Testing began by fixing the ignition timing at 8° BTDC and the equivalence ratio at 1.0. The intake temperature was allowed to stabilize at a steady-state temperature of 10°C (50°F). The manifold air pressure was set in the Woodward OH1.2 ECU to 131 kPa (19 psia). The manifold air pressure was then incrementally increased by 13.78 kPa (2 psi) until the engine began to knock (IV<0.87). At each manifold air pressure increment, engine data and emissions measurements were taken. Upon encountering knock (IV<0.87), the boost was reduced by 6.89 kpa (1psi) and recorded as the knock limited point of operation. Emissions measurements and engine data for this point were also taken. The same procedure was conducted for two other intake air temperatures, 26°C (80°F) and 54.4°C (130°F).

One of the goals of this project was to record data at the same BMEP with an intake air temperature of 10°C (50°F) and 54.4°C (130°F). The manifold air pressure before the inception of knock at 10°C (50°F) was predicted to be higher than at 54.4°C (130°F). To obtain a test point with the same BMEP, a lower manifold air pressure at both 10°C (50°F) and 54.4°C (130°F) would need to be achieved. The turbocharger on this engine was selected to run as a lean burn engine, which produces higher manifold air pressures than a stoichiometric engine at the same torque output. With the wastegate fully open, the lower boost pressure could not be achieved at the 10°C (50°F) intake air temperature. To achieve the lower manifold air pressure, air was bled off of the intake system through a valve in the separator. The valve position on the separator was held constant for all steady-state testing.

Transient Testing - In order to conduct the transient tests, engine parameters (ignition timing, equivalence ratio, and boost pressure) were reset to their base values as a lean burn engine. Similarly, the valve located on the separator was completely closed. The engine's fuel supply was changed back to a typical CNG composition (96.42% Methane, 1.71% Ethane, 0.19% Propane, 0.12% Butane, 1.01% $CO_2$, 0.54% $N_2$, and 0.01% Other). Once again, the engine was allowed to obtain a steady-state manifold air temperature of 10°C (50°F) at 1600 rpm, WOT. The engine was then taken down to an idle and the dynamometer speed control was set to 1600 rpm. After achieving a steady idle speed, the throttle was immediately opened. While the dynamometer used is designed for steady-state operation, the internal water gap can absorb a significant amount of power generated during this acceleration from idle to 1600 rpm. During this test, the OH 1.2 was used to collect second by second engine parameter data. The recorded parameters include engine speed, manifold air pressure, manifold air temperature and $NO_x$ emissions. The test was performed once more using 54.4°C (130°F) intake air to provide a performance comparison.

## RESULTS

This section presents the main findings of this project. Among these findings are the resultant knock limited operation of the engine at various intake air temperatures. Emissions results from the progressive increase in boost pressure are presented to provide insight into the effect of reduced temperature boost air on various exhaust emissions. Also included in this section are the transient test results and heat exchanger effectiveness results.

ENGINE RESULTS - The following figures present results of the testing. Testing was conducted at a constant equivalence ratio of 1.0 and ignition timing of 8°BTDC. Figure 2 depicts the BMEP achieved through the progressive increase in manifold air pressure. As expected, BMEP increases with increased manifold air pressure, and lower intake air temperatures have a higher BMEP at the same manifold air pressure due to the higher air density. The highest pressure associated with each curve indicates the point just before the engine starts to knock. Figure 3 represents the knock limited BMEP as a function of the intake air temperature. As intake air temperature decreases, the knock limited BMEP increases. The knock limited BMEP increase from 54.4°C (130°F) to 10°C (50°F) is 30.7% (501 kPa). Preliminary testing indicated that at lean burn conditions, the knock limited BMEP increase is also approximately 30% (490 kPa). Figure 4 shows the brake specific fuel consumption (BSFC) at constant BMEP as a function of

temperature. The BSFC rises slightly with increasing intake air temperature.

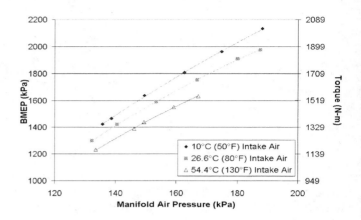

Figure 2. Torque Increase Due to Increased Manifold Air Pressure

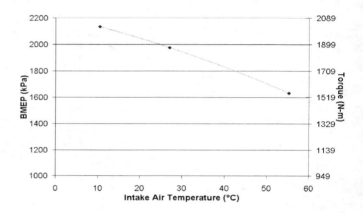

Figure 3. Knock Limited BMEP as a Function of Intake Air Temperature

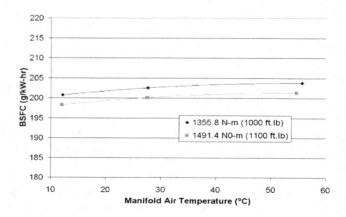

Figure 4. BSFC as a Function of Intake Air Temperature

14

EMISSIONS RESULTS - This section presents an emission comparison between the various intake air temperature points tested. Data presented in the following graphs are points obtained at the approximately the same torque value. Testing was conducted at a constant equivalence ratio of 1.0 and ignition timing of 8°BTDC. The only parameter altered from point to point is the intake air temperature and the boost pressure that was required to achieve the specified torque. There are two lines on each graph, one representing a torque of approximately 1355.8 N-m (1000 ft.lbs) and the other representing a torque of approximately 1491.3 N-m (1100 ft.lbs). Figure 5 depicts the uncorrected brake specific oxide of nitrogen (BSNO$_x$) emissions as a function of intake temperature. At the engine operating condition selected, the uncorrected BSNO$_x$ emissions are not significantly different with varying intake air temperature. The BSNO$_x$ emissions are expected to decrease with a lower intake air temperature, but due to intake air humidity differences, the uncorrected BSNO$_x$ emissions did not change. At the 54.4°C (130°F) intake air temperature, the humidity was the same as the ambient humidity, which was approximately 12.4 g$_{H2O}$/kg$_{air}$. At the 10°C (50°F) intake air temperature, the relative humidity was 100% at the outlet of the secondary heat exchanger, which was approximately 6.0 g$_{H2O}$/kg$_{air}$. Figure 6 depicts the corrected BSNO$_x$ emissions. The correction factor was applied for the humidity after the second heat exchanger. The corrected BSNO$_x$ emissions are 17.8% (3.4 g/kW-hr) lower with the cooler intake air temperature. A similar representation is given in Figures 7 and 8 for the brake specific hydrocarbon (BSHC) and brake specific carbon monoxide (BSCO) emissions. Figure 7 shows that at the engine operating condition selected, the brake specific HC emissions are 39.9% (0.5 g/kW-hr) lower with the cooler intake air temperature. The lower humidity may have an effect on lowering the HC emissions. Figure 8 shows that the brake specific CO emissions are 7.5% (0.2 g/kW-hr) higher with the cooler intake air temperature. The colder intake air, and therefore colder cylinder wall temperatures, may not allow the hydrocarbons to fully oxidize, creating higher CO emissions.

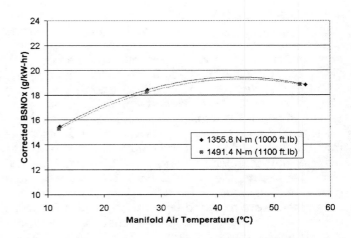

Figure 6. Corrected BSNOx Emissions as a Function of Intake Air Temperature (corrected for humidity at intake manifold)

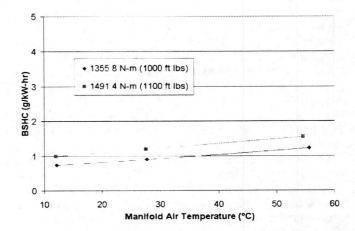

Figure 7. BSHC Emissions as a Function of Intake Air Temperature

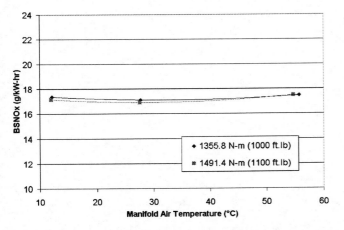

Figure 5. BSNOx Emissions as a Function of Intake Air Temperature (constant ambient humidity)

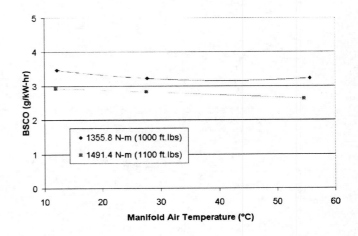

Figure 8. BSCO Production as a Function of Intake Air Temperature

TRANSIENT TEST RESULT – This section describes the results of the transient test performed. The engine calibration was restored to the original lean burn configuration, with varying equivalence ratio and ignition timing based on engine speed and manifold air pressure. This test was performed to provide a transient response comparison of the engine at two different intake air temperatures. The engine was first allowed to reach steady-state at 1600 rpm and WOT. The engine was then brought down to idle. Once a stable idle speed was established, the engine was accelerated at WOT to 1600 rpm. The first temperature tested was at a steady-state intake air temperature of 10°C (50°F). The engine parameters were recorded on a second by second basis during this test. The same test was then performed with a steady-state intake air temperature of 54.4°C (130°F). Figures 9 and 10 represent the time history of the cooled and uncooled intake air system transient response. Figure 9 shows the manifold air pressure reached the steady-state value approximately 11.5% (1.21 seconds) quicker at the lower temperature, and Figure 10 shows the engine speed reached a steady-state value approximately 15.3% (1.43 seconds) quicker at the lower temperature. The quicker time to spool up the turbocharger is a result of the lower humidity and higher density air with the lower intake air temperature. Figure 11 shows the realtime NOx emissions of the cooled and uncooled air system as measured by the in-exhaust NOx sensor during the transient tests. The average uncorrected NOx emissions for the cooled intake air is 825 ppm, whereas the average uncorrected NOx for the uncooled intake air is 752 ppm. The average corrected NOx emissions for the cooled intake air is 710 ppm, whereas the average corrected NOx for the uncooled intake air is 783 ppm. Figure 12 shows the manifold air temperature of the cooled and uncooled air system. The manifold air temperature starts above the 1600 rpm, WOT steady-state temperature due to heating of the air by the engine. The airflow at idle is slow enough, such that heat transfer from the engine raises the temperature of the air in the manifold. The temperature eventually reached the 1600 rpm WOT steady state temperature.

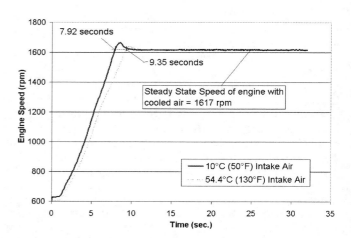

Figure 10. Engine Speed Time History

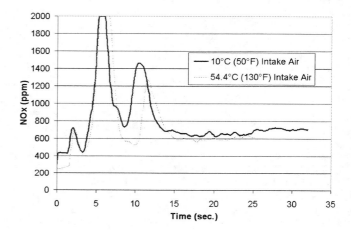

Figure 11. NOx Emissions Time History

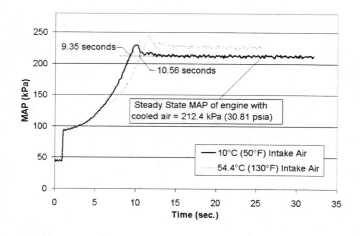

Figure 9. Manifold Air Pressure Time History

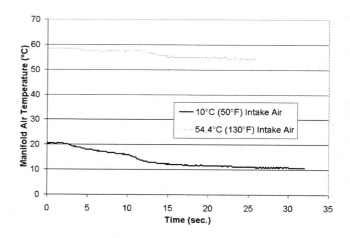

Figure 12. Manifold Air Temperature Time History

16

HEAT EXCHANGER PERFORMANCE - This section briefly addresses the effectiveness of the simulated LNG cooler to lower the intake air to the desired temperature. During testing, the inlet and outlet temperatures of each heat exchanger were recorded. Figure 13 presents the temperature profile of the intake air system as it travels through the various heat exchangers. Inlet and outlet temperatures of the separator were not recorded. For the 10°C (50°F) test point, the water to the primary heat exchanger was at full flow, and the temperature of the water to the second heat exchanger was controlled. For the 26°C (80°F) test point, the temperature of the water to the second heat exchanger was set to 18.3°C (65°F), which was the highest set point available from the chiller, and the water flow to the primary heat exchanger was controlled. For the 54.4°C (130°F) test point, the chiller was turned off so that the second heat exchanger was transferring heat to ambient conditions, and the water flow to the primary heat exchanger was controlled.

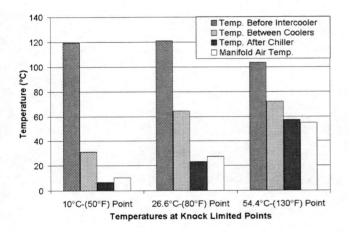

Figure 13. Heat Exchanger Effectiveness

## CONCLUSIONS AND RECOMMENDATIONS

After successful completion of this project's objectives, a number of conclusions and recommendations can be made. This section presents a summarized listing of such conclusions and recommendations.

1. The engine would not knock before reaching the safe operating limit (2155.75 Nm, 1590 ft-lb) of the Mack E7G engine when using 10°C (50°F) intake air. At this point the engine parameters were set to provide a lean air to fuel ratio (φ=0.67) and full boost (289.57 kPa, 42 psia).

2. In order to bring the engine into a knocking condition when using 10°C (50°F) intake air, the equivalence ratio was increased to provide stoichiometric fueling (φ=1.0).

3. To help protect the engine, a water separator should be placed in the air intake system to remove the condensed water that is generated by the cold intake air temperatures.

4. The reduced intake air temperature allowed an increase of 30.7% (501 kPa) in knock limited BMEP. (This comparison is made between the 54.4°C and the 10°C points at an ignition timing of 8° BTDC and an equivalence ratio of 1.0.)

5. At constant BMEP, there was a 0.6% (0.1 g/kW-hr) reduction in the uncorrected BSNO$_x$ emissions with the reduced intake air temperature, and a 17.8% (3.4 g/kW-hr) reduction in corrected BSNO$_x$ emissions. (This comparison is made between the 54.4°C and the 10°C points at the 1355.8 N-m torque point)

6. At constant BMEP, there was a 39.9% (0.5 g/kW-hr) reduction in the BSHC emissions with the reduced intake air temperature. The reduction in HC emissions is possible due to the lower humidity. (This comparison is made between the 54.4°C and the 10°C points at the 1355.8 N-m torque point)

7. At constant BMEP, there was a 7.5% (0.2 g/kW-hr) increase in the BSCO emissions with the reduced intake air temperature. (This comparison is made between the 54.4°C and the 10°C points at the 1355.8 N-m torque point)

8. Transient response showed improvements in both the time to achieve a set speed as well as the time to achieve a set manifold air pressure. There was a 1.21-second decrease in the time required to reach 212.42 kPa (30.81 psia) when using the chilled air. Similarly there was a 1.43-second decrease in the time required to obtain an engine speed of 1617 rpm. This corresponds to an 11.5% decrease and 15.3% decrease respectively in the time to these set points. This is due to the lower humidity and higher density with the chilled air.

9. An "LNG air cooler" on a vehicle would allow for higher BMEP due to knock limitations. It could also allow for stoichiometric operation at the current power level of the Mack E7G engine, which would otherwise not be possible without other methods to reduce the tendency to knock.

## ACKNOWLEDGMENTS

The work described in this paper was conducted by Southwest Research Institute under contract from Advanced Technology Management. The authors would like to acknowledge several people whose contributions were instrumental in the completion of this project: Jim Wegrzyn of Brookhaven National Laboratory who provided the funding for this project through the U.S. Department of Energy Office of Heavy Vehicle

Technologies, Ken Murphy of Mack Truck Inc. for allowing the E7G engine to be used in this project, and Steve Almaraz and Richard Waren of Southwest Research whose excellent technical skills were greatly responsible for the success of this project.

## REFERENCES

1. Meyer, R.C., Shahed, S.M., "An Intake Charge Cooling System for Application to Diesel, Gasoline and Natural Gas Engines," SAE Paper 910420, Society of Automotive Engineers, 1991
2. Cole, J., J. Chiu, and J. Bartel, "Development of the Mack E7 Natural Gas-Fueled Engine," The 5th Biennial INGV International Conference and Exhibition on Natural Gas Vehicles, Kuala Lumpur, Malaysia, 1996.

2001-01-3683

# Preliminary Investigation on the Viability of 1,3-Dioxolane as an Alternative to MTBE in Reformulated Gasoline

**Patrick J. Flynn, Mku Thaddeus Ityokumbul and Andre´ L. Boehman**
The Energy Institute, The Pennsylvania State University

## ABSTRACT

An experimental investigation was conducted to determine the efficiency of 1,3-dioxolane as an alternative oxygenate to MTBE in Reformulated Gasoline. In the investigation, the effect of adding 1,3 dioxolane on octane rating was evaluated. The octane number of the fuels was determined using a Waukesha single cylinder, 4-stroke cycle, 2-valve, CFR F-2U octane rating unit. Certified 87 octane gasoline was used as the base fuel which 1,3-dioxolane was added at specific volumetric proportions. Iso-octane (100 octane number) and N-heptane (0 octane number) are primary reference fuels that were blended at volumetric proportions to produce a reference base of known octane number. The reference base fuel of known octane number was used for comparison of knock tendency to the test fuels under the ASTM D 2699 (Research) and ASTM D 2700 (Motor) methods of testing. Cost analyses were conducted to determine and show the volume addition comparisons for MTBE, ethanol, and dioxolane to comply with current RFG regulations. Pricing of dioxolane was also evaluated to find out what the maximum production cost can be to still be competitive in the oxygenate market. A theoretical price increase of the wholesale sales of gasoline due to dioxolane blending for RFG production was also included. The results obtained from the octane rating analysis show that as volume percentage of dioxolane was added to the base fuel there was a steady increase in octane number. Furthermore, the theoretical, and experimental results clearly demonstrate that 1,3-dioxolane has the ability to be an effective oxygenate for RFG production. The cost analyses demonstrated that dioxolane will be able to decrease volume addition by 1.05 and 1.5 % respectively compared to ethanol to achieve compliance with the current 2% and 2.7% oxygen RFG standards. Furthermore, because of this volume decrease, dioxolane will be able to be sold wholesale up to $1.78/gallon to yield equivalent blending costs as ethanol to achieve the 2% oxygen RFG standard. The results, and cost analyses, were both discussed in detail to demonstrate why 1,3-dioxolane could be a viable alternative to MTBE and recommendations were given as to what should be accomplished before dioxolane's viability can be determined.

## INTRODUCTION

Air pollution control strategies in the automotive industry have been evolving since the 1940's when California first hypothesized that automobile emissions and air pollution problems were linked [1,2]. This hypothesis, reached through studies early that decade, encouraged further research that soon prompted regulatory acts. Ever since motor vehicle emission regulations and control strategies were instituted in the United States, allowable levels have been progressively reduced due to excessive automobile fleet growth and continuing environmental degradation. Constant improvements of vehicle technology and fuel formulations have been the result of progressive change in regulations set forth in the Clean Air Act and its amendments [1]. Motor vehicles emit high quantities of carbon monoxide (CO) during the cold months of the year because of incomplete burning of fuel. During cold winter periods the combustion of fuel slows and enables unburned fuel to escape through the exhaust. Unburned fuels in motor vehicle exhaust streams are a very large source of CO emissions that tend to enhance the formation of smog. Carbon monoxide is a precursor of tropospheric ozone which in turn reacts with oxides of nitrogen to form urban smog [3,4,5,6].

Methyl-tertiary-butyl-ether (MTBE), a well known oxygenate is presently used to enable compliance with the Clean Air Act Amendments by reducing urban smog formation resulting from tropospheric ozone. Starting in 1992, with the cooperation of the U.S. EPA, petroleum companies began adding MTBE to gasoline to improve combustion and decrease harmful carbon monoxide emissions during the colder months of the year. The goal was to improve air quality in urban areas. Because of MTBE's powerful emission reduction effects it has become a major component of U.S. gasoline. It is used primarily in the production of reformulated gasoline (RFG) mandated in the Clean Air Act Amendments of 1990 and similar regulations set forth by the California Air Resources Board (CARB) [7,10]. The use of RFG has been responsible for substantial reductions of smog forming pollutants resulting in enhanced urban air quality. These guidelines will be discussed in a later section of the paper. The EPA estimates that smog-forming pollutants are being reduced annually by at least 105,000 tons. Furthermore, fuel analysis data from industry for compliance purposes show that emission reductions from the RFG program have exceeded the program requirements every year since the program started in 1995 [8]

## BACKGROUND

CLEAN AIR ACT - The Clean Air Act is the comprehensive federal law that regulates air emissions from area, stationary, and mobile sources. The Act began in 1963 under the direction of the Department of Health and was structured to strengthen state and local air pollution control efforts. In 1965 the act was amended and specifically authorized the writing of national standards for emissions from all new motor vehicles sold in the United States. In 1968 control of motor vehicle exhaust emissions began. Gasoline vehicles were required to be equipped with closed crankcase ventilation systems, and tailpipe emissions were limited to 275 ppm of HC and 1.5% of CO. Beginning in 1970 the federal government authorized the U.S. Environmental Protection Agency to establish National Ambient Air Quality Standards (NAAQS) to be amended to the Clean Air Act. The goal was to set and achieve NAAQS in every state by 1975. By 1971 the maximum allowable tail-pipe emissions for light-duty spark-ignited (SI) engines was set to be 2.2 g/mile of total hydrocarbons (THC), and 23 g/mile of carbon monoxide CO. By 1977 maximum allowable tail-pipe emissions standards for light-duty SI engines plummeted to 0.41 g/mile for THC, and 3.4 g/mile for CO, while adding control for oxides of nitrogen (NOx) to not exceed 1.0 g/mile. Although these 1977 emissions standards brought about significant improvements in our nation's air quality, the urban air pollution problems of ozone (smog), and carbon monoxide, still persisted. The 1990 Clean Air Act Amendments were created to combat the problem of urban smog. Target dates were established for a two-phase motor vehicle emissions reduction plan that is as follows:

Phase I: 0.25 g/mile HC, 0.40 g/mile NOx, and 3.4 g/mile CO had to be reached by 1994. The CO emission standard remained the same because this standard was not met under the 1977 standards.

Phase II: Based on air quality in 1997, phase II required further reductions of emission rates to 0.125 g/mile for HC, 0.2 g/mile for NOx, and 1.7g/mole for CO. These standards have to be met by 2003.

The amendments also call for regulation of the composition of gasoline in specified non-attainment areas by limiting the volumetric fractions of benzene, requiring a specific amount of oxygen in the fuel (2% by wt. during ozone non-attainment periods and 2.7% by wt. during CO non-attainment periods). Ozone non-attainment usually occurs during the summer months, and CO non-attainment occurs during the winter months [8]. The above specifications are listed under the Reformulated Gasoline Program and are accomplished by the addition of the following oxygenates (oxygen-rich compounds):

- Methyl-tertiary-butyl-ether (MTBE)
- Ethanol
- Ethyl-tertiary-butyl-ether (ETBE)
- Methanol

Selected cities have different time periods for mandated RFG use depending on climate and smog formation. For example most major cities in the state of California use RFG all year round due to the existence of both warm and cold climates and extensive smog formation.

The RFG program was implemented in 1995 under the Clean Air Act of 1990 to enable complete combustion of fuel. This program has resulted in substantial reductions in the emissions of a number of air pollutants from motor vehicles, most notably volatile organic compounds and carbon monoxide (precursors of tropospheric ozone), and mobile-source air toxics (benzene, 1,3-butadiene, and others) [9]. Over 85% of RFG used in the U.S. contains the oxygenate (MTBE) and approximately 8% contains Ethanol. The remaining 7% of RFG contains a combination of the other oxygenates (Methanol, ETBE, etc) [11]. Even though no one oxygenate reduces emissions better than another, MTBE has been the most preferred oxygenate by refiners because of its low cost, ease of production, and blending characteristics.

MTBE CONCERN - A growing number of studies have detected MTBE in ground water and surface water throughout the country; in some instances these contaminated waters are sources of drinking water [10,11,12,13,14]. Low levels of MTBE can make drinking water supplies unpalatable due to its low taste and odor threshold [12] The majority of the human health-related research conducted to date on MTBE has focused almost exclusively on effects associated with the inhalation of the chemical. Results found that when research animals inhaled high levels of MTBE, they have developed cancers or experienced other non-cancerous health effects. To date, independent expert review groups who have assessed MTBE inhalation health risks (e.g., "Interagency Assessment of Oxygenated Fuels") have not concluded that the use of MTBE-oxygenated gasoline poses an imminent threat to public health. However, researchers have limited data about the health effects caused if a person swallows (ingests) MTBE. EPA's Office of Water has concluded that the data available is not adequate to estimate potential health risks of MTBE at low exposure levels in drinking water, however the data supports the conclusion that MTBE is a potential human carcinogen at high doses. Recent studies by EPA and other researchers are expected to help determine more precisely the potential for health effects from MTBE in drinking water. The EPA concluded in it's drinking water advisory that at this time there is little likelihood that MTBE in drinking water will cause adverse health effects at concentrations between 20 and 40 ppb or below [11,12,13,14]. However, the agency is very concerned and has been continuing to follow closely the increasing detections of MTBE in the ground and surface water supplies throughout the nation.

While most MTBE detections typically occur below the advisory levels of 20 to 40 ppb, there have been many instances of contamination at much higher levels of potential health concern [12,13,14]. In areas where higher concentrations were detected people are complaining of respiratory irritation, nervousness, dizziness, lightheadedness, nausea, insomnia, headache, watering of eyes, skin rash, increased heart rate, long lasting colds, etc. However, while these symptoms have not been proven to be related to MTBE; they are the basis of continuing water quality assessments and other studies of MTBE.

THE CONTROVERSY - Concern about MTBE encroaching into water supplies quickly spread outside of California to other parts of the country. In 1990, the EPA set up a blue-ribbon panel made up of industrial executives representing a wide range of concern and support. The panel recommended that the use of MTBE should be reduced because it threatens water quality [11].

The ethanol industry claims that it is the answer to continued air quality enhancement and the solution to the MTBE problems. In March 2000, Carol Browner, the EPA administrator, held a press conference with the Secretary of Agriculture to put pressure on Congress to ban MTBE and to substitute an ethanol mandate for the RFG provisions of the Clean Air Act. The oil industries disagreed. [7].

Despite the differences of opinion related to MTBE's future, all parties agreed that the advancement of clean-burning fuels must be continued if MTBE is to be eliminated. The California Air Resource Board adjusted its Phase-III Gasoline regulations to make it easier to use ethanol and to help spur the elimination of MTBE by the end of 2002. Bills were brought to Congress varying from proposing the elimination of the oxygen standard in RFG, to a complete ban on MTBE, to a mandate that all gasoline contain a certain amount of ethanol. As of now, Congress has taken no final action, and the RFG program has changed very little. MTBE usage is higher than ever, market prices are rising, and refineries are finding it difficult to find an adequate supply. This is due in part because; rising MTBE usage in Europe has reduced foreign shipments to the U.S., limiting available supplies. Even though MTBE is still being widely used throughout the country, momentum to phase out or reduce usage of MTBE is growing. Therefore, it appears desirable that research should continue to investigate for possible oxygenate alternatives that have limited logistical and technical disadvantages [7].

## OBJECTIVE

The objective of this thesis is to determine the efficiency of using 1,3-dioxolane as an alternative oxygenate to MTBE in RFG production. The hypothesis of this thesis is that dioxolane is a possible alternative to MTBE and is worthy of further studies to prove its viability. The intention of this investigation is to determine if 1,3-dioxolane (CAS # 646-06-0) can contribute to solving MTBE problems, while minimizing specific environmental, logistical, and technical problems that may result when the use of MTBE is either eliminated or reduced.

## EXPERIMENTAL

ENGINE AND TEST EQUIPMENT - An experimental investigation was conducted to evaluate the effect of 1,3-dioxolane addition on octane rating. The octane number of the fuels was determined using a Waukesha single cylinder, 4-stroke cycle, 2-valve, CFR F-2U octane rating unit. The CFR F-2U Rating Unit is a combination of the F-1 and F-2 Rating Units. The F-2U Rating Unit provides the capability to switch between Research and Motor Methods of testing, with only minor equipment adjustments. Engine specifications and parameters are presented in Table 1.

To enable this switch from motor and research methods, the engine flywheel is connected by belt to an electric synchronous-reluctance type motor that is mounted on a sliding base connected to the bedplate. This specially designed electric motor acts to start the engine, absorb the power output of the engine, and maintain constant engine speed. Sample test fuels and reference fuels are introduced to any of four carburetor bowls mounted on a carburetor assembly. The carburetor features a horizontal venturi with fuel entering through a single vertical jet; while the carburetor bowls and associated float chambers provide the ability to vary the fuel-air ratio.

The rating unit is equipped with an intake mixture manifold and an electric mixture manifold heater assembly, between the carburetor and the engine cylinder, to raise the temperature of the air-fuel mixture to the proper American Society of Testing Materials (ASTM) specifications. The ignition coil is connected to a spark plug installed in the side of the engine cylinder. A solid-state ignition control assembly unit charges a capacitor, which is discharged through the primary winding of an ignition coil when actuated by a timing sensor on the engine camshaft. Spark timing is therefore adjusted to specific Research/Motor Method specifications.

**Table 1.** Engine Parameters and Characteristics

| ITEM | DESCRIPTION |
|------|-------------|
| Bore Diameter | 3.250 inches |
| Stroke | 4.500 inches |
| Compression Ratio (C.R.) | Adjustable 4:1 to 18:1 by Cranked Worm Shaft - Worm Wheel Drive in Cylinder Clamping Sleeve |
| Displacement | 37.33 cubic inches |
| Camshaft Overlap | 5 degrees |
| Minimum Speed | 600 rpm (relates to mild on road operation) |
| Maximum Speed | 900 rpm (relates to severe on road operation) |
| Valve Mechanism | Open Rocker Assembly With Linkage for Constant Valve Clearance as C.R. Changes |
| Ignition | Electronically Triggered Condenser Discharge Through Coil to Spark Plug |
| Ignition Timing | Research Method: 13 degrees BTDC Motor Method: Variable as Cylinder Height (C.R.) is changed (base setting 26 degrees BTDC at 5:1 C.R.) |

Engine inlet air and mixture temperatures are established using electric heaters. The fuel-air mixture temperature changes relative to the test method and therefore is controlled by temperature control instrumentation. The engine's coolant system, intake air system, and surge tank were designed and installed separately. The cooling water system taps into the utility water line of the building. The utility water is then filtered and passed through a reverse osmosis system and introduced to the engine at 1.5 GPM. The intake air system conditions the existing air to the proper humidity through an air humidity control system (refrigeration unit). Exhaust gasses are directed through a flexible exhaust manifold to an exhaust surge tank for dispersal of gasses outside the laboratory through PVC piping.

Instrumentation for measuring the intensity of knock during testing includes a pickup, and transducer installed in the top of the engine cylinder, a detonation meter to condition the knock signals, and a 0-100-division knock meter mounted on the instrument panel to display the signal.

The rating unit was also coupled with a Pentium PC-based data acquisition system (DAS). The DAS is based on a National Instruments GPIB 488.2 card, which enable communications with an HP3497 data sampling/processing unit. Graphical programming and data acquisition and instrumentation software (Labview),

developed by National Instruments was used to program a Virtual Instrument that enabled the transformation of thermocouple voltage signals into a temperature display. This enabled continuous monitoring of heated and non-heated intake air temperatures, exhaust inlet and outlet temperatures, and coolant temperatures. A spark-plug-mounted pressure transducer was also installed in the engine to produce a pressure voltage signal. The pressure signal was then passed through a charge amplifier and displayed on an oscilloscope to show knock behavior.

TESTING AND OPERATION - Octane number measurements were conducted under standard testing procedures set forth by the American Society of Testing Materials. The standard test methods used for Research Octane Number and Motor Octane Number of spark-ignition engine fuel were ASTM D 2699 and ASTM D 2700, consecutively. Testing was performed consecutively from 1% dioxolane to 4% dioxolane by volume starting with the measurement of Research Octane Number. Measurements of Motor Octane Number were conducted after all Research Octane Number ratings were established for each percentage of dioxolane added and the rating unit was reconfigured and recalibrated to Motor Method standards. Standard test method operating conditions are presented in Table 2.

FUELS - Certified 87 octane gasoline was used as the base fuel in which 1,3-dioxolane was added at specific volumetric proportions. Some of the dioxolane's properties supplied by the Environmental Protection Agency (EPA) are presented in Table 3. Iso-octane (100 octane number) and N-heptane (0 octane number) are primary reference fuels that were blended at volumetric proportions to produce a reference base of known octane number. The reference base fuel of known octane number was used for comparison of knock tendency to the test fuels under the ASTM D 2699 (Research) and ASTM D 2700 (Motor) methods of testing. dioxolane was obtained from Ferro Corporation and the primary reference fuels were purchased from Phillips 66 Petroleum Company.

Dioxolane is not a common industrial chemical. It is a stable reaction product of ethylene glycol and formaldehyde. Its primary uses are:

- Co-monomer for the manufacture of polyacetals and other polymers.
- Solvent for chemical reactions (including inorganic salts).
- Stabilizer for halogenated organic solvents.
- As a starting material or reagent for organic synthesis

Among all current uses of dioxolane, 95% or more of its current production is consumed in either the production of polyacetals or as starting material for the production of pharmaceuticals. The remaining 5% is used as an industrial solvent. Furthermore, there are currently no non-industrialized applications of the compound, thus all human exposure occurs in industrial settings.

**Table 2.** Operating Conditions

| OPERATING CONDITION | RESEARCH METHOD | MOTOR METHOD |
|---|---|---|
| Engine Speed | 600 +/- 6 rpm | 900 +/- 9 rpm |
| Fuel-Air Mixture Temperature | | 300 +/- 2 °F |
| Spark Timing | Constant 13 degrees BTDC | Variable (base setting 26 degrees BTDC) |
| Spark Plug Gap | 0.020 +/- 0.005 in. | 0.020 +/- 0.005 in. |
| Oil Pressure | 25 to 30 psi | 25 to 30 psi |
| Oil Temperature | 135 +/- 15 °F | 135 +/- 15 °F |
| Coolant Temperature | 212 +/- 3 °F | 212 +/- 3 °F |
| Inlet-Air Temperature | 83 +/- 2 °F | 100 +/- 5 °F |
| Intake-Air Humidity | 25 to 30 grains of $H_2O$/lb dry air | 25 to 30 grains $H_2O$/lb dry air |
| Valve Clearances | 0.008 +/- 0.001 in. | 0.008 +/- 0.001 in. |

**Table 3.** Properties of 1,3- Dioxolane

| Empirical Formula | $C_3H_6O_2$ |
|---|---|
| Molecular Weight | 74.09 |
| Flash Point | 35 °F |
| Density/Specific Gravity | 1.0600 @ 20 °C |
| Bulk Density | 8.2 lb/gal @ 20 °C |
| Solubility | Soluble in water, alcohol, ether, acetone |
| Melting Point | Negative 95 °C |
| Boiling Point | 78 °C @ 765mm Hg |
| Vapor Pressure | 70mm @ 20 °C |
| Partition Coefficient | Log Ko/w= -37 |

## RESULTS AND DISCUSSION

OCTANE RATING - The specific octane number rating outcomes for the base fuel and all four oxygenated test fuels are shown in Figure 1. The test fuel's results are shown over each blending percentage (1-4%). Research, Motor, and Average Octane Number (R+M/2)

characterize all the results. The error bars shown in the graphs correspond to the confidence intervals related to relative error in each measured quantity of test and reference fuels blended. These intervals were established on the basis of how the measurements would affect the final octane number rating result, according to methods described by Moffat [15]. Similar confidence intervals from the actual rating error of octane number by the Waukesha CFR-F2U rating unit were not established due to insufficient data regarding the limits of error for various instrumentation for which the octane number is determined.

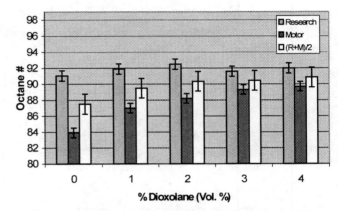

**Figure 1.** Variation of octane # with dioxolane addition by volume.

The deviations among Research, Motor, and Average Octane Numbers for the base fuels were 8.9, 2.7, and 4.5 percent, respectively. The deviations at one percent oxygenate addition were 8.4, 2.6, and 1.5 percent, respectively. The differences between Research and Motor Octane Number deviation was observed to arise from four factors:

1. Differences in engine speed
2. Differences in spark timing
3. Differences in air-inlet temperature
4. Detonation meter control setting variability (controls rate of sampling of cylinder pressure to provide a voltage signal to send to the knock meter and display knock intensity)

The variations of octane number with the percent oxygenate addition to the base fuel is shown in Figures 2-7. The trends show how octane number was enhanced as each percentage of oxygenates was added. The first three graphs (Figures 2-4) show these trends for Research Octane Number (RON), Motor Octane Number (MON), and the Average Octane Number (R+M/2) (AON). The last three graphs (Figures 5-7) show these trends for the change in Research Octane Number, Motor Octane Number, and Average Octane Number. The error bars shown in these graphs are the same confidence intervals shown in Figure 1. However, they represent a visual explanation of how the trend lines could look if 100 percent accuracy and precision were achievable.

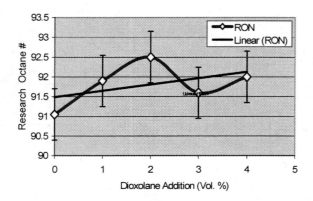

Figure 2. Variation of Research Octane Number with Dioxolane Addition by Volume.

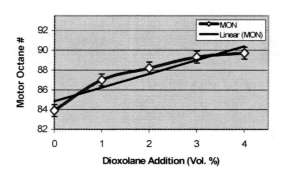

Figure 3. Variation of Motor Octane Number with Dioxolane Addition by Volume.

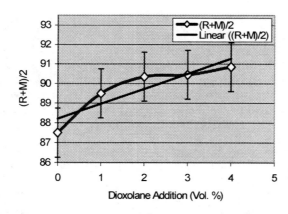

Figure 4. Variation of Average Octane Number with Dioxolane Addition by Volume.

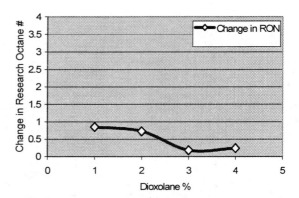

Figure 5. Change in Research Octane Number with Dioxolane Addition by Volume.

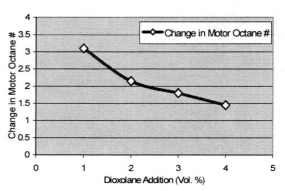

Figure 6. Change in Motor Octane Number with Dioxolane Addition by Volume.

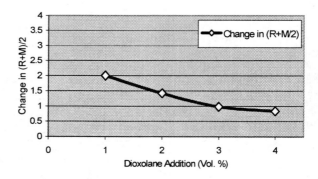

Figure 7. Change in Average Octane Number with Dioxolane Addition by Volume.

COST - Cost is very important in considering dioxolane as a viable oxygenate to replace MTBE. Can 1,3-dioxolane potentially be competitive in price with current oxygenates used to meet RFG standards? If dioxolane were to replace MTBE, production capacity would be achievable. Dioxolane is manufactured from formaldehyde, and ethylene glycol, both very abundant and low in cost. Also, current production of dioxolane is on a very small scale due to its limited uses. Small production volumes leave the dioxolane industry opportunity for expansion in production. Environmentally, dioxolane has been known to biodegrade in about a year in water that is not exposed to air, and due to its volatility characteristics photodegredades within about 10 to 30 hours in water that is exposed to aerobic environments. Furthermore, dioxolane is relatively non-toxic to humans. The Toxicology and Regulatory Affairs show that the most sensitive targeted organ is the bone marrow. Studies show that at concentrations above 300 ppm, the production of white blood cells is reduced. Therefore, at high concentrations dioxolane may cause a person to become anemic [16]. This is well above the majority of known MTBE detections. However, this was only found through direct oral and inhalation routes.

Surface water is not a challenge due to dioxolane's photodegredation characteristics. Ground water contamination is a topic that still needs to be addressed and proven. Studies by the Toxicology and Regulatory Affairs in November of 2000 show that dioxolane does biodegrade within a year [16].

Cost is a large issue at the present stage of dioxolane research. Therefore the results of the analyses carried out are shown below. Analysis #1 shows the volume addition comparisons for MTBE, ethanol, and dioxolane to comply with current RFG regulations. Dioxolane will be able to decrease volume addition by 1.05 and 1.5 % respectively compared to ethanol to achieve compliance with the current RFG standard. This analysis was performed to demonstrate possible cost savings based on the volume difference of dioxolane compared to MTBE, and ethanol. Cost savings will occur if the price of dioxolane is below the comparable price to blend ethanol since ethanol is currently the most desirable oxygenate alternative. Analysis #2 shows the price that dioxolane has to be in order to compete in the oxygenate market. The prices shown are the dioxolane cost equivalent to blend ethanol. If dioxolane's wholesale cost is equivalent or below the prices shown than dioxolane will be competitive. Analysis #3 represents the price increase of RFG for each oxygenate.

Analysis #1 – This analysis is based on the 2.0 and 2.7% oxygen by weight standard in RFG production. This is a comparison between the volumetric contributions of MTBE, ethanol, and dioxolane required for the production of RFG respectively, and the costs associated with it.

> **MTBE:** In industry today, the bulk of RFG production is an MTBE blend. An MTBE RFG blend consists of 11 or 14% MTBE by volume is blended at the refinery in order for petroleum company's to comply with the RFG standards of 2.0 and 2.7% oxygen by weight respectively.

Molecular Formula: $CH_3O(CH_3)$
Molecular Weight: 88g/mole

**Molecular Structure:**

$$-CH_3-O-\underset{\underset{CH_3}{|}}{\overset{\overset{CH_3}{|}}{C}}-CH_3$$

**Ethanol:** Molecular Formula: $CH_3CH_2OH$
Molecular Weight: 46 g/mol
Molecular Weight of $O_2$: 16 g/mol

$$\text{Therefore: } \frac{16 \text{ g O}}{46 \text{ g } CH_3CH_2OH} = 0.347$$
$$= 35\% \ O_2 \text{ by Wt.}(\surd)$$

$$\frac{2.0\% \text{ RFG } O_2 \text{ Standard}}{0.347} = \begin{array}{l} 5.7\% \text{ by Vol.} \\ \text{of Ethanol} \end{array}$$

**Explained:**
This volume percent of ethanol needs to be added to a base gasoline to produce an ethanol blended RFG, compliant at 2% $O_2$ by Wt.

$$\frac{2.7\% \text{ RFG } O_2 \text{ Standard}}{0.347} = \begin{array}{l} 7.78\% \text{ by} \\ \text{Vol. of} \\ \text{ethanol} \end{array}$$

**Explained:**
This volume percent of ethanol needs to be added to a base gasoline to produce an ethanol blended RFG, compliant at 2.7% $O_2$ by Wt.

**1,3-Dioxolane:** Molecular Formula: $C_3H_6O_2$
Molecular Weight: 74.09 g/mol
Molecular Weight of $O_2$: 32 g/mol
Molecular Structure:

$$\text{Therefore } \frac{32 \text{g } O_2}{74 \text{g } C_3H_6O_2} = 0.43 \ (\surd)$$
$$= 43\% \ O_2 \text{ by Wt.}$$

$$\frac{2.0\% \text{ RFG } O_2 \text{ Standard}}{0.43} = \begin{array}{l} 4.65\% \text{ by Vol. of} \\ \text{Dioxolane} \end{array}$$

**Explained:**
This volume percent of dioxolane needs to be added to a base gasoline to produce a dioxolane blended RFG, compliant at 2.0% $O_2$ by Wt.

$$\frac{2.7\% \text{ RFG } O_2 \text{ Standard}}{0.43} = \begin{array}{l} 6.28\% \text{ by Vol. of} \\ \text{Dioxolane} \end{array}$$

**Explained:**
**This volume of dioxolane needs to be added to a base gasoline to produce a dioxolane blended RFG, compliant at 2.7% $O_2$ by Wt.**

Analysis #2 - This cost comparison is based on two variables: (A) current prices of ethanol and MTBE as of April 5th, 2001 [17](B) the estimated costs of ethanol and MTBE relative to the national yearly average price of gasoline [17]. The second variable was established because ethanol, MTBE, and gasoline prices change daily. These variables were established to effectively provide a basis for evaluating a competitive price of 1,3-dioxolane relative to the volume addition needed to meet the RFG standards. Both comparisons evaluate the competitive price of dioxolane assuming it is the same price that it costs to blend ethanol for 2% oxygen compliance. Ethanol averages approximately (40-45 cents/gallon) above the cost of gasoline and MTBE averages approximately (50-55 cents/gallon) above the cost of gasoline [18]. This assumption is established because ethanol is currently the most competitive alternative to MTBE in RFG production.

**(A) Variables:**   Ethanol   = $1.30/gallon
                    MTBE   = $1.43/gallon
                    1,3-dioxolane = X

**Facts:**   <u>RFG Standards</u>
- 2% $O_2$ by Wt($O_3$ Non-attainment)
- 2.7% $O_2$ by Wt(CO Non-attainment)

<u>Volume % Oxygenate Addition for Compliance:</u>

- **(2%):** MTBE = 11% by volume
  Ethanol = 5.7% by volume
  1,3-dioxolane = 4.65% by volume

- **(2.7%):** MTBE = 14% by volume
  Ethanol = 7.7% by volume
  1,3-dioxolane = 6.28% by volume

**MTBE:**
11($1.43/gallon) = 0.1573 cents/gallon
      (1 gallon, 2% $O_2$ blend)
14($1.43/gallon) = 0.2002 cents/gallon
      (1 gallon, 2.7% $O_2$ blend)

**Ethanol:**
5.7($1.30/gallon) = <u>0.0741 cents/gallon</u>
      (1 gallon, 2% $O_2$ blend)
7.7($1.30/gallon) = 0.1001 cents/gallon
      (1 gallon, 2.7% $O_2$ blend)

**Competitive Dioxolane Price:**

$\frac{4.65}{100}$ (X) = 0.0741 cents/gallon

$$\boxed{X = \$1.59/\text{gallon}} \quad (\checkmark)$$

**Explained:**
**Therefore, dioxolane would be competitive at $1.59 or below based on the above assumptions**

**(B) Variables:**   Gasoline   = $1.00/gallon
                    Ethanol   = $1.45/gallon
                    MTBE   = $1.55/gallon
                    1,3-dioxolane = X
**Facts:**   <u>RFG Standards</u>
- 2% Oxygen by Wt($O_3$ Non-attainment)
- 2.7% Oxygen by Wt(CO Nonattainment)

<u>Volume % Oxygenate Addition for Compliance:</u>

- **(2%):** MTBE = 11% by volume
  Ethanol = 5.7% by volume
  1,3-dioxolane = 4.65% by volume

- **(2.7%):** MTBE = 14% by volume
  Ethanol = 7.7% by volume
  1,3-dioxolane = 6.28% by volume

**MTBE:** 11($1.55/gallon) = 0.1705 cents/gallon
      (1 gallon, 2% O blend)
14($1.55/gallon) = 0.2170 cents/gallon
      (1 gallon, 2.7% O blend),

**Ethanol:** 5.7($1.45/gallon) = <u>0.0827 cents/gallon</u>
      (1 gallon, 2% O blend)
7.7($1.45/gallon) = 0.1165 cents/gallon
      (1 gallon, 2.7% O blend)

**Competitive Dioxolane Price:**

$\frac{4.65}{100}$ (X) = 0.0827 cents/gallon

$$\boxed{(X) = \$1.78/\text{gallon}} \quad (\checkmark)$$

**Explained:**
**Therefore, dioxolane would be competitive at $1.78 or below based on the above assumptions**

Analysis #3 - Note: This cost analysis is based on the change in production costs of a 1% dioxolane blend to produce an 89-octane gasoline for a 1-gallon sample. The addition of 1% dioxolane provides approximately a 2-octane unit increase.

**Baseline gasoline price: $1.00 gallon**

**1% dioxolane addition by volume: .01 gallons**

**Batch: 0.99 gallons 87-octane fuel**
**0.01 gallons 1,3-dioxolane**

## Based on Current Prices:

**1,3-dioxolane: .01 gallons dioxolane x $\frac{\$1.59}{gallon} =$**

= $0.0159 cents ⌐ Based on Case A
Analysis #2 ¬

**87-octane fuel:**
$0.90 cents per gallon ➔ $0.90 x .99 gallons
= $0.891 cents

**1 gallon of dioxolane blend = 0.9069 cents**

## Explained:
**Therefore dioxolane will increase current prices of gasoline by 0.0069 cents/gallon**     (√)

## Based on Theoretical Prices:

**1,3-dioxolane: .01 gallons dioxolane x $\frac{\$1.78}{gallon} =$**

= $0.0178 cents ⌐ Based on Case B
in Analysis # 2 ¬

**87-octane fuel:**
$1.00 per gallon ➔ $1.00 x .99 gallons
= $0.99 cents

**1 gallon of dioxolane blend = $1.0078**

## Explained:
**Therefore dioxolane will increase current prices of gasoline by 0.0078 cents/gallon**   (√)

The investigation that was carried out and presented here is just the first step toward determining if 1,3-dioxolane is a viable alternative to MTBE? Even though this investigation cannot be fully support the hypothesis of this thesis with the data presented, this work is intended to be the basis for further studies of its viability.

The results obtained from the octane rating analysis show that as volume percentage of dioxolane was added to the base fuel there was a steady increase in octane number. We see an initial accelerated boost in octane number with a 1% addition, then a continued gradual increase with further addition until the trend slowly linearizes.

This octane rating analysis clearly demonstrates the ability of 1,3 dioxolane to be an effective oxygenate for RFG production. Trends from this analysis are congruent with other oxygenate investigations, and reinforce the validity of the tests [19,20].

Dioxolane's cost comparison to MTBE appears promising, as production volumes increase costs are expected to decrease in the future. Future cost analyses of dioxolane should be concentrated on the following scenario: what the wholesale cost of dioxolane will be once production volumes increase to a point where dioxolane can support demand.

The oxygen standard for RFG also becomes an issue related to cost. The oxygen evaluation of dioxolane shows that in order to comply with these standards, 4.65 and 6.28% by volume would need to be added to obtain 2.0 and 2.7 oxygen by weight in RFG compared to MTBEs 11 and 14%. This shows a 42% reduction in volume additions to achieve the same standard, and a similar reduction in the quantity of MTBE that would need to be produced.

The U.S. Government seems to be leaning toward ethanol as the permanent future alternative. Ethanol has production volume problems, economical problems, similar environmental mobility problems as MTBE, and toxicity problems.

Ethanol is a by-product of corn. This fact raises the question, will we be able to continually meet the United State's agricultural needs for food production, livestock feed, as well as the increased ethanol demand? What will happen during dry seasons or famine while these demands continue to increase? Furthermore, the oil and gas journal states that if ethanol were to replace MTBE, current production would have to triple requiring major investments, vast increases in subsidies, and great faith on the part of investors that their crops would not meet the same fate as did the MTBE merchant market. Also, oil industries say that in order for ethanol to become a viable means of eliminating or reducing the use of MTBE, workable policies for maximizing ethanol use would have to include reducing the oxygen content requirement in RFG. An oxygen content of approximately 1.3 wt% would be needed in order to incorporate a 100% switch over, or if the use of MTBE is to be continued then it would have to pick up the difference of 0.7 and 1.4 wt % respectively [7]. Is this correct? If ethanol is decided to be the alternative/replacement, or foothold to MTBE reductions, then these arguments need to be addressed before it is mandated for use.

## RECOMMENDATIONS AND CONCLUSIONS

Momentum to phase out MTBE seems to be unstoppable. California is scheduled to eliminate use of

MTBE by the end of 2002 and other states will also be likely to reduce or eliminate the use of MTBE by 2005, depending on future oxygenate requirements. The U.S. government appears to support the renewable fuels industry and seems likely to encourage or even demand the use of ethanol. However despite the significant logistical and technical disadvantages to using ethanol, government seems to be favoring this approach because it feels using ethanol is better for air quality than the elimination of oxygen requirements and the use of oxygenates altogether [7,2121]. Oxygenates comparatively differ in refining characteristics, cost, and production not in air quality benefits.

Addition of 1,3-dioxolane to 87 gasoline resulted in a clearly demonstrated enhancement of Research, Motor, and Average Octane Number. However, the beneficial effect of dioxolane gradually starts to diminish with volume additions above 2%. From the data obtained in this investigation, dioxolane appears to be a good candidate for the replacement of MTBE in RFG.

To fully answer the objective of this thesis, the following are recommended:

1. Octane enhancement testing should be continued for volume additions above 4%. Determinations need to be made to see if octane enhancement starts to diminish after a certain elevated volume percentage is reached.
2. Emissions testing should be conducted for dioxolane use in RFG for comparison with MTBE trends from the past.
3. Cost evaluations of dioxolane should be conducted for conditions as if it became the alternative to MTBE and production volumes rose to meet demand.
4. Toxicology studies should be continued to answer any questions as to if dioxolane could have the potential of being a carcinogen. Studies need to establish whether dioxolane is characterized as a known human carcinogen, as a possible human carcinogen, or not a carcinogen at all, in order for its viability to be determined.
5. Re-evaluations of dioxolanes mobility in the subsurface and ground water also will need to be established to reinforce studies done in the past.
6. Dioxolane's octane number should be established just for commercial and industrial information.
7. Other physical and chemical properties of the blended fuels should be studied, such as density, viscosity, calorific value, and Ried vapor pressure (RVP).

Therefore, in conclusion research should press on to fully characterize dioxolane's viability as a replacement of MTBE in RFG production.

## REFERENCES

[1]    F. Black, "Automobile Emissions," Encyclopedia of Energy Technology and the Environment, New York: Wiley, c(1995).

[2]    J. Robert Mondt, " Cleaner Cars, The History and Technology of Emission Control Since the 1960's," SAE International, Society of Automotive Engineers, Inc. (2000).

[3]    E.F Obert, "Internal Combustion Engineers and Air Pollution," Ch.10, pp.342-381, 42 U.SF. s/s 7401 et seq, Harper Collins Publishers, New York, N.Y., (1973).

[4]    S. Poulopoulos and C. Philippopoulus, "Influence of MTBE Addition into Gasoline on Automotive Exhaust Emissions," Atmospheric Environment, 34 4781-4786 (2000).

[5]    J. Heywood, Internal Combustion Engine Fundamentals, McGraw-Hill, New York (1988).

[6]    M. Walsh, and A. Virginia, "Air Pollution: Automobile", Encyclopedia of Energy Technology and the Environment, New York: Wiley, c(1995).

[7]    D. Miller, "MTBE Faces an Uncertain Future," Oil and Gas Journal, July 10, 2000.

[8]    E.P.A., "Clean Air Act," http://www.epa.gov/regions/defs/html/caa.htm, s/s 7401 et seq. (1970).

[9]    EPA, "The Blue Ribbon Panel on Oxygenates in Gasoline – Executive Summary and Recommendations", Document # 202-564-9674, http://www.epa.gov/oms/consumer/fuels/oxypanel/ bluerebb.htmUS.EPA,202-564-9674., (July 1999).

[10]   P. Squillace, J. Zogorski, W. Wilber, and C. Price, "A Preliminary Assessment of the Occurrence and Possible Sources of MTBE in Groundwater of the United States," U.S. Geological Survey Open File Report 95-456, 1993-1994.

[11]   B. Simmons, "MTBE – Where is it, and How Did it Get There," http://www.grac.org/spring96/chemist.html, Feb 1996.

[12]   P. Squillace, "MTBE in the nations Ground Water, National Water Quality Assessment (NAWQA) Program Results." Http://sd.water.usgs.gov/nawga/vocns/brp-pjs- handout.html U.S. Geological Survey, April 29, 1999.

[13]   United States Geological Survey, "Occurrence Of the Gasoline Oxygenate MTBE and BTEX Compounds in Urban Storm Water in the U.S.", Water-Resources Investigations Report 96-4145, (1999).

[14]   EPA, "Methyl Tertiary Butyl Ether (MTBE), Drinking Water", Http://www.epa.gov/mtbe/water.htm

[15]   R. Moffat, "Describing the Uncertainties in Experimental results", Professor of Mechanical Engineering, Stanford University, Stanford, California, Essevier Science Publishing Co. Inc. New York, (1988).

[16]   Toxicology and Regulatory Affairs, "USEPA HPV Challenge Program Submission," submitted by: Dioxolane Manufactures Consortium Members,

Ferro Corporation and Ticona (November 20, 2000).

[17] Oxyfuel News, A Chemical Weekly Associates Publications, Hart Publications Inc., Editor: Rachel Gantz, (March 2001, week #1).

[18] Ron V. Lamberty, (personal communication) Director Ethanol Market Development, American Coalition for Ethanol, (March, 2001).

[19] H. Hess, J. Szybist, A. Boehman, P. J. A. Tijm, and F. J. Walker, "Impact of Oxygenated Fuel on Diesel Engine Performance and Emissions," Proc. 35th Natl. Heat Trans. Conf., ASME Paper No. NHTC01-11462, June 2001, Anaheim, CA (2001).

[20] T. Litzinger, M. Stoner, H. Hess, and A. Boehman, "Effects of Oxygenated Blending Compounds on Emissions from Turbo Charged Direct Injection Diesel Engine," *Int. J. Engine Res.*, **1**, 57-70 (2000).

[21] California Environmental Protection Agengy, "Basis For Waiver Of The Federal Reformulated Gasoline Requirement For Year-Round Oxygenated Gasoline In California", http://www.calepa.ca.gov/Programs/mtbe/O2Waiver.htm, (2001).

# Engine Control System Architecture for Bi-Fuel Vehicles

**Allen Dobryden, Peter Kuechler and John Lapetz**
Ford Motor Company

## ABSTRACT

This paper describes a new architecture for a bi-fuel vehicle engine control system, which can reduce system cost while improving function. The proposed architecture uses a modified (dual parameter) PCM strategy to control operation on both fuels, with a simpler additional module to drive fuel injectors and interface to other alternative fuel components. It is shown that this architecture results in improved fuel control and lower tailpipe emissions compared to typical aftermarket systems. Impact on the development process and base vehicle wiring are minimized.

## INTRODUCTION

Bi-fuel vehicles are desirable in some applications to take advantage of alternative fuels such as compressed natural gas (CNG) and liquefied petroleum gas (LPG), but also address fueling station availability problems. Such vehicles have a price premium due to the added fuel system, and are also subject to increasingly strict emissions and OBD requirements.

Most current bi-fuel vehicles use either an add-on control system with its own fuel control module, or a modified powertrain control module (PCM) capable of driving two sets of fuel injectors.

There are various aftermarket systems available for conversion of production vehicles to run on alternative fuels. These controllers can be expensive, difficult to install and calibrate, and generally have inadequate fuel control to meet increasingly lower emission standards. In addition, operation of on-board diagnostics during alternative fuel operation of the vehicle will soon be a requirement. This will require extensive improvements to the aftermarket control systems to work properly.

Producing a new bi-fuel PCM is cost prohibitive, given the generally lower production volumes of alternative fuel vehicles. The constraints posed by the low volumes, stringent emissions targets, and OBD requirements caused us to develop a new approach to bi-fuel powertrain control.

The proposed architecture uses a modified (dual parameter) PCM strategy to control operation on both fuels, with a simpler additional module (Bi-Fuel Interface Module, or "BFIM") to drive fuel injectors and interface to other alternative fuel components. A communication protocol will be outlined which is used to transfer information between the PCM and BFIM. Features of the BFIM module will be described, as well as results from emission testing with the new system.

## PROJECT GOALS

The project for development of the new architecture had four major goals:

1. Improved Fuel Control – Improved control is required to meet lower tailpipe emissions targets.

2. Universality - As alternative fuel vehicles are generally a low volume product, the architecture should be universal to allow maximum application across product lines. Different vehicle product lines typically have different PCM types due to varying system requirements and the desire to minimize PCM cost. Any solution which involves a unique alternative fuel PCM is costly and would need to be duplicated for each vehicle type. The applications envisioned involve from 4 to 10 cylinder engines, and both CNG and LPG fuels. The ability to drive various numbers and types of fuel injectors is required.

   An added bonus would be the ability to use the new module on dedicated (i.e. single fuel) alternative fuel vehicles as well.

3. Common algorithms, process, and tools – leverage of the highly developed gasoline fuel control

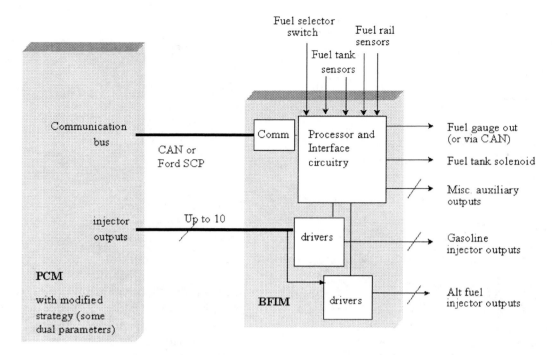

**Figure 1: Bi-Fuel System Architecture**

strategies is desirable from a functional standpoint as well as for process benefits. The use of standard PCM calibration procedures and tools for development takes advantage of existing knowledge base and resources. The use of standard dealer tools for field service and reflashing are critical to acceptance of the technology.

4. Ease of installation - installation of alternative fuel components and wiring may be done at the vehicle assembly plant, or at an outside modifier. Solutions which require sharing of sensors are undesirable (from a functional standpoint as well). Underhood packaging for the control module is preferred as most of the interfacing is to underhood components.

## SYSTEM ARCHITECTURE

Figure 1 shows a block diagram of the control system architecture. The PCM hardware is an unmodified gasoline PCM. The PCM software is modified to add alternative fuel functionality.

The BFIM module adds any control system hardware functionality required by the addition of the alternative fuel system. Its main functions are:

- Provide high current driver circuits for a set of multi-port alt fuel injectors

- Provide driver circuits for alternative fuel tank and rail isolation solenoids

- Measure alternative fuel tank level
- Measure alternative fuel rail pressure and temperature; convey to PCM
- Determine operating fuel, based on driver desire and alt fuel level; convey to PCM
- Perform diagnostics for service and OBD

Gasoline injector driver circuits are duplicated in the BFIM to allow a direct means of turning them off in alt fuel mode.

All fuel metering and other engine control strategy resides in the PCM. The BFIM merely acts as an interface device. The PCM strategy is derived from the base gasoline version of the strategy. A number of parameter tables are added for use during alternative fuel operation. Certain other functions unique to alternative fuel operation are added, such as gaseous fuel rail pressure compensation.

## MODULE INTERACTION

At ignition key-on, the BFIM initializes itself, turns off both sets of injector drivers, and reads the state of the fuel selector switch in the cockpit. The following steps then take place, with communication between the modules taking place over the vehicle's CAN or SCP bus:

## Alt Fuel Operation Selected

1. The BFIM requests the PCM to start in alt fuel (i.e. refer to it's internal set of alt fuel parameters for fueling).

2. If the PCM agrees to the request, the BFIM enables the set of alt fuel injector drivers and opens the alt fuel isolation solenoids.

3. From then on during the trip, the BFIM periodically transmits alt fuel rail pressure and temperature to the PCM, which is used as part of the fuel pulsewidth calculation.

4. If the BFIM detects a low alt fuel level on start, it requests the PCM to switch to gasoline operation, disables the alt fuel injector drivers, enables the gasoline injector drivers and fuel pump, and closes the alt fuel isolation solenoids.

## Gasoline Operation Selected

1. The BFIM requests the PCM to start in gasoline (i.e. refer to its internal set of gasoline parameters for fueling).

2. If the PCM agrees to the request, the BFIM enables the set of gasoline injector drivers and fuel pump.

## Fuel Switchover

1. If the BFIM detects that the alternative fuel is depleted (by detecting tank level or a drop in fuel rail pressure), the BFIM requests the PCM to revert to gasoline operation.

2. If the PCM agrees to the request, the BFIM enables the set of gasoline injector drivers and fuel pump.

# PCM STRATEGY

Extensive experience running dedicated CNG and LPG vehicles on Ford gasoline based control systems provided a good basis for determining the requirements for an integrated bi-fuel control system. The CNG and LPG fuel control requirements differ from gasoline in a number of areas including combustion characteristics, volumetric efficiency, octane rating, and wall wetting/fuel puddle behavior.

- The reduced flame speed of CNG and LPG relative to gasoline requires a more advanced MBT spark timing than that of gasoline. The flame temperatures also differ from gasoline, driving differences in exhaust temperature.

- The gaseous nature of the injected CNG and LPG displaces a significant amount of the air that would normally be ingested for a given throttle position, reducing the volumetric efficiency. This requires re-calibration of control features that rely on the relationship between throttle position and airflow.

- Octane rating of CNG and LPG is much higher than gasoline allowing operation at MBT spark for almost all operating conditions.

- The gaseous nature of CNG and LPG does not have a characteristic of wetting the port, valve, and combustion chamber with fuel as with gasoline, negating the need for transient fuel compensation. Additionally, the gaseous fuels mix very well with air and have significantly less uncombusted, or lost, fuel that must be accounted for.

- Some additional changes were required to allow for characterization of the CNG and LPG fuel injectors and for fuel density compensation of the gaseous fuel.

Changes to the control system are summarized in the following table:

| Control Feature | Changes | Justification |
|---|---|---|
| Crank fuel | Duplicate tables for alt fuel and gasoline | Combustion characteristics |
| Injector Characterization | Duplicate injector slopes and offsets; Fuel rail density compensation for gaseous fuels | |
| Spark | Duplicate MBT and borderline spark tables | Flame speed, octane rating |
| Transient Fuel | Duplicate tables | Negligible wall wetting |
| Manifold filling | Duplicate inferred load, man. pressure functions | Volumetric efficiency |
| Open Loop Fuel | Duplicate tables | Flame speed and lost fuel |
| Catalyst Temperature | Duplicate tables | Flame temp., vol. efficiency |
| Emissions Bias | Duplicate tables | Catalytic behavior |

| | | |
|---|---|---|
| Idle Control | Duplicate tables | Volumetric efficiency |
| BP Model | Duplicate tables | Volumetric efficiency |
| Adaptive Fuel | Split tables | Allow separate storage of CNG and gasoline adaptive fuel terms |

## BFIM FEATURES

1. Alternative Fuel Injector Drivers – the BFIM module has 10 peak-and-hold driver circuits capable of up to 6 amp peak current and 1.5 amp hold current operation. The module also has the capability of running the driver in a saturation mode for a programmable length of time. This mode may be desired for cold start operation for example.

2. Gasoline Fuel Injector Drivers – there are channels for 10 saturation-type drivers (nominal 1 amp current).

3. Other Control Outputs – the module has a number of auxiliary outputs for driving various solenoids associated with an alternative fuel system. It also has the capability of PWM outputs, which can be used for functions such as speed control of LPG pumps.

4. Sensor Inputs – there are analog inputs for sensors such as fuel rail temperature and pressure, fuel tank temperature and pressure for CNG tanks, and fuel tank float sensors for LPG tanks.

5. Communication Interfaces – the module has a CAN interface, an ISO 9141 interface, and a Ford SCP interface.

6. Diagnostics – the module is designed to provide both on-board diagnostics related to emissions function, and service-related diagnostics for dealer service use. Open/short detection on both types of injector driver is available, as well as out-of-range sensor detection. In addition, the module is capable of being re-programmed in the field by standard dealer service tools.

## SYSTEM PERFORMANCE

### A/F Standard Deviation

Improved fuel control was one of the primary objectives of the new control system architecture. A method used to quantify the fuel control of various fuel systems was to record lambda over the emissions drive cycle and calculate the standard deviation. Five powertrain control systems were compared:

- MY00 Ford Focus -- aftermarket CNG control
- MY00 Ford Focus -- EEC/BFIM CNG control
- MY02 Ford Transit -- EEC/BFIM LPG control
- MY02 Ford bi-fuel CNG F250
- MY02 Ford F-250 -- EEC/BFIM CNG control

A summary of the results is given in the table below:

| Control system | Powertrain | Average lambda | Standard deviation |
|---|---|---|---|
| EEC/BFIM Focus | 5spd Manual 2.0L | 0.9855 | 0.0240 |
| Aftermarket Focus | 5spd Manual 2.0L | 0.9719 | 0.0456 |
| EEC/BFIM LPG Transit | 2.3L 5spd Manual | 0.9831 | 0.0215 |
| MY02 Bi-Fuel F250 | 5.4L 4spd Auto | 1.0081 | 0.0447 |
| MY02 EEC/BFIM F250 | 5.4L 4spd Auto | 1.0047 | 0.0160 |

The following chart illustrates typical air-fuel ratio excursions during tip-ins and tip-outs on a MY02 Ford Bi-Fuel F250 versus an EEC/BFIM-based vehicle:

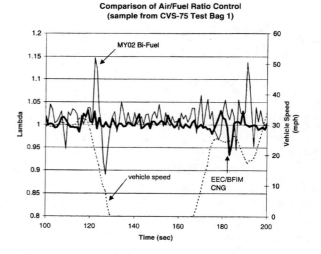

This data clearly illustrates that the fuel control provided by the highly developed EEC control strategy is substantially better than that provided by previous systems. Subjective drive evaluations have rated the BFIM-based system to have improved driveability.

**Tailpipe Emissions**

A further measure of fuel control can be found in the examination of tailpipe emissions. Previous CNG and LPG fuel systems have historically had difficulty meeting the increasingly stringent emission standards, in many cases having trouble meeting the performance of their highly refined gasoline counterparts. Ford OEM CNG vehicles have set many benchmarks in emissions performance and this was a goal of the new bi-fuel control system. The vehicle used as a benchmark was the MY02 Ford bi-fuel CNG F250. An identical vehicle was outfitted with an EEC/BFIM control system and both were tested for 4K emissions performance.

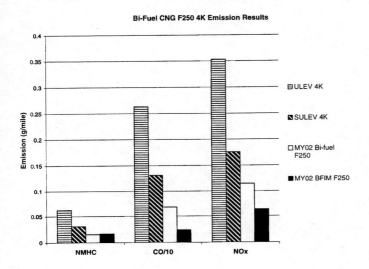

The EEC/BFIM control system performed much better in both CO and NOx with equivalent NMHC and the results were well below SULEV 4K requirements. Additionally, test-to-test variability is greatly reduced compared to the existing system.

## CONCLUSION

Adopting this architecture has enabled us to meet our goals for the next generation of bi-fuel vehicles.

The BFIM-based fuel control system has been shown to have improved fuel control. The use of gasoline fuel control algorithms tuned for the gaseous fuel, with multi-port injection, gives optimal performance. The system was shown to exhibit lower tailpipe emissions than a comparable aftermarket system.

The powertrain calibration development process is nearly identical to the equivalent gasoline vehicle process. Existing calibration tools can be used to manipulate the bi-fuel parameters. Calibration procedures developed for gasoline engines can be easily adapted for use with alternative fuels.

The architecture allows easy integration with the vehicle system. In addition to an overlay wiring harness for the added fuel system, incorporation of the module with an existing vehicle wiring harness requires only re-routing the PCM injector outputs and tying into the communication bus. No existing sensors need to be shared or duplicated.

## ACKNOWLEDGMENTS

The authors would like to thank Cosworth Technology, Inc. for its role in development of the BFIM module.

## CONTACT

For questions related to this paper, please contact Al Dobryden (email: adobryde@ford.com).

## DEFINITIONS, ACRONYMS, ABBREVIATIONS

**BFIM:** Bi-Fuel Interface Module.

**CNG:** Compressed Natural Gas

**EEC:** Electronic Engine Control – refers to engine control module.

**LPG:** Liquefied Petroleum Gas

**MBT:** Mean Best Torque

2002-01-1704

# Isotopic Tracing of Bio-Derived Carbon from Ethanol-in-Diesel Blends in the Emissions of a Diesel Engine

**Bruce A. Buchholz**
Center for Accelerator Mass Spectrometry, Lawrence Livermore National Laboratory

**A. S. (Ed) Cheng and Robert W. Dibble**
Mechanical Engineering Department, University of California, Berkeley

## ABSTRACT

The addition of oxygenates to diesel fuel reduces particulate emissions, but the mechanisms responsible for the reductions are not well understood. Measurement of particulate matter (PM), unburned hydrocarbons (HC), and carbon monoxide (CO) are routine, but determining the origin of the carbon atoms that make up these undesired emissions is difficult. The sub-attomole ($<6 \times 10^5$ atoms) sensitivity of accelerator mass spectrometry (AMS) for measuring carbon-14 ($^{14}C$) allows tracing the carbon atoms from specific fuel components to soot or gaseous emissions. Radioactive materials are not required because contemporary carbon (e.g., ethanol from grain) has 1000 times more $^{14}C$ than petroleum-derived fuels. The specificity of the $^{14}C$ tracer and the sensitivity of AMS were exploited to investigate the relative contribution to diesel engine PM, CO, and $CO_2$ from ethanol and diesel fractions of blended fuels. The test engine, a 1993 Cummins B5.9 diesel rated at 175 hp 2500 rpm, was operated at steady-state conditions of 1600 rpm and 210 ft-lbs. PM was collected on quartz filters following a mini-dilution tunnel. The limited solubility of ethanol in diesel fuel required either an emulsifier (Span 85) or cosolvent (n-butanol) to prepare 10, 20, and 40% ethanol-in-diesel blends. An ignition improver, di-tert-butyl peroxide (DTBP), was added to give all blends the same autoignition properties as the baseline diesel. PM was separated into volatile and non-volatile organic fractions (VOF and NVOF) for AMS analysis. The homogeneous cosolvent blends were more effective in reducing total PM mass, but the heterogeneous emulsified blends yielded larger VOF that are easily treated by exhaust catalysts. Ethanol derived carbon tended to reside in the NVOF, especially for the cosolvent blends.

## INTRODUCTION

Environmental and human health concerns over emissions from internal combustion engines continue to bring about increasingly stringent emissions standards and drive research into the use of non-conventional, cleaner-burning fuels. For compression-ignition (diesel) engines, oxygenated fuels have been shown to dramatically reduce particulate matter (PM) while also improving or maintaining acceptable levels of other regulated emissions ($NO_x$, HC and CO) [1-14]. The mechanisms through which oxygenates reduce PM, however, are unclear. In addition to changes in combustion chemistry, the influence of thermophysical properties on fuel injection and fuel-air mixing can play a significant role.

Researchers in the mid-1980s labeled fuel components with $^{14}C$ and traced the radioisotope to PM or soot from a diesel engine [15] or diffusion flame [15,16] using a decay-counting technique. These experiments required special radioactive test facilities to contain the large amounts of volatile radioactive compounds needed for decay counting and housing a radioactive engine. In addition to generating a significant amount of radioactive and mixed wastes, high level radioactive tracing can never be used in a realistic engine environment. The high sensitivity of accelerator mass spectrometry (AMS) allows the specificity of the $^{14}C$ atom to be used while avoiding radioactive waste issues. Furthermore, AMS detection permits tracing with road vehicles in conventional dynamometer facilities or on the open road.

Radioisotopes are specific and distinctive because they are extremely rare in natural materials. A radioisotope-labeled compound has a very high abundance-to-background ratio in natural systems, but poor signal-to-noise in the isotope detector may obscure this property. For example, the natural level abundance of $^{14}C$ due to cosmic radiation is 1.2 parts in $10^{12}$. The rare stable isotope of carbon, $^{13}C$, is naturally 1.1%. A one ppm concentration of a $^{13}C$-labelled compound (assume formula weight 200 g/mol) will change the $^{13}C$ concentration by only 0.3 per 1000, measurable under good conditions using an excellent mass spectrometer. The same material labeled with $^{14}C$ changes the concentration of that isotope in a contemporary biological sample by a factor of 3 million. Efficient detection of radioisotopes is a key to using this specificity.

Short-lived isotopes can be efficiently detected by their decay but produce high radiation hazards in the laboratory. Radioisotopes that have longer half lives (e.g., $^{14}C$ half life = 5730 y) are inefficiently detected by measuring decays. Measuring only 0.1% of the $^{14}C$ decays in a sample requires uninterrupted counting for 8.3 years ( 0.1% x 5730 y / ln(2) ). The sensitivity and specificity of the radioisotope label are wasted in detecting decays. Accelerator mass spectrometry (AMS) is an isotope-ratio measurement technique developed in the late 1970s as a powerful tool for tracing long-lived radioisotopes in chronometry in the earth sciences and archaeology [17]. Samples prepared for $^{14}C$ analysis are combusted to $CO_2$ and then reduced to graphite for use in the AMS ion source. The technique counts individual nuclei rather than waiting for their radioactive decay, allowing measurement of more than 100 $^{14}C$ samples per day.

The contemporary quantity of $^{14}C$ in living things (1.2 parts in $10^{12}$ or 110 fmol $^{14}C$ / g C) is highly elevated compared to the quantity of $^{14}C$ in petroleum-derived products. Accordingly, components of bio-derived fuels contain elevated $^{14}C$ as compared to fossil fuels. This isotopic elevation is sufficient to trace the fate of bio-derived fuel components in the emissions of an engine without the use of radioactive materials. The complications of licensing and radioactive waste disposal are completely avoided. Some petroleum-derived components can be synthesized from biological sources. If synthesis of a fuel component from biologically-derived source material is not feasible, another approach is to purchase $^{14}C$-labeled material and dilute it with petroleum-derived material to yield a contemporary level of $^{14}C$. In each case, the virtual absence of $^{14}C$ in petroleum based fuels gives a very low $^{14}C$ background that makes this approach to tracing fuel components practical.

Regulatory pressure to significantly reduce the particulate emissions from heavy-, medium- and light-duty diesel engines is driving research into understanding mechanisms of soot formation. If mechanisms are understood, then combustion modeling can be used to evaluate possible changes in fuel formulation and suggest possible fuel components that can improve combustion and reduce PM and other emissions. The current combustion paradigm assumes that large molecules break down into small components and then build up again during soot formation. If all fuel components are broken down into 1- or 2-carbon species, then there should be no difference in the contribution of carbon from aromatics, alkanes or oxygenates in the PM emissions. AMS allows the labeling of specific carbon atoms within fuel

components, tracing the carbon atoms, and testing this combustion modeling paradigm.

Volatile and non-volatile organic fractions (VOF, NVOF) in the PM can be further separated. Researchers at Southwest Research Institute (SwRI) showed that the addition of the oxygenate dimethoxymethane to diesel fuel not only reduced total PM emissions, but also increased the proportion of VOF in the PM [18,19]. The VOF of the PM can be oxidized in the exhaust stream to further decrease PM. A fuel formulation that significantly shifts PM to the VOF can be more valuable in reducing emissions than a variation that merely drops total PM.

The methods described below for tracing fuel components in the emissions of diesel engines can be applied to any engine or combustion system. Any molecule containing carbon can be labeled with $^{14}C$. Techniques for measuring the $^{14}C$ concentrations by AMS are straightforward and routine. Knowing the chemical identity and carbon inventory of a sample is the key to exploiting the power of $^{14}C$-AMS.

Meeting future emission requirements will likely require modification of existing diesel fuel and exhaust treatment to reduce $NO_x$ and oxidize PM. The decisions which drive approaches to satisfying pending emission regulations must balance engine and emission performance, cost, fuel compatibility with the existing vehicle fleet, national security, and fuel supply. If reducing dependence on imported oil becomes a priority, use of bio-derived fuels will probably be required. Bio-diesel and ethanol are obvious options, each with its own features. Bio-diesel is immediately usable but production capacity is limited. Ethanol has a much larger production capacity, but it is not a good diesel fuel. The ease of tracing bio-derived ethanol against a petroleum background and the simplicity of the molecule influenced our decision to demonstrate the power of AMS to trace fuel component carbon in PM and gaseous emission from a variety of ethanol-in-diesel blends.

## OBJECTIVE

Our goal was to demonstrate methods for tracing bio-derived carbon from fuel components in the emissions from a diesel engine. In addition to demonstrating methods, we present preliminary data showing the partitioning of labeled oxygenates in various fractions of particulate and gaseous emissions.

## FACILITIES

ENGINE FACILITY – Emission samples were collected from a 1993 Cummins B5.9 engine at the Combustion Analysis Laboratory at the University of California at Berkeley (UCB). Figure 1 shows the engine installed at UCB and detailed specifications are listed in Table 1. Fuel injection is achieved with a Bosch P7100 PE type inline pump capable of injection pressures of up to 115 MPa. No modifications were made to the engine or fuel injection system to optimize for operation on the test fuels.

**Figure 1.** Cummins B5.9 Engine installed at UC Berkeley's Combustion Analysis Laboratory.

**Table 1.** Cummins B5.9 engine specifications

| Model year | 1993 |
|---|---|
| Displacement | 5.88 liters (359 in$^3$) |
| Configuration | 6 cylinder inline |
| Bore | 102 mm (4.02 in) |
| Stroke | 120 mm (4.72 in) |
| Compression ratio | 17.6:1 |
| Horsepower rating | 175 hp @ 2500 rpm |
| Torque rating | 420 ft-lb @ 1600 rpm |
| Aspiration | turbocharged and aftercooled |
| Injection timing | 11.5° BTDC |

Fuel consumption was determined using a Micro Motion R025 coriolis flow meter. Modifications to the fuel return system were made to eliminate fuel return to the storage tank. The return line is instead routed back into the fuel delivery line and a shell-and-tube heat exchanger was installed to prevent overheating of the fuel in the short-circuited system.

Gaseous emissions were monitored using Horiba gas analyzers as listed in Table 2. Measurements of PM were made via a mini-dilution tunnel designed and constructed by UCB.

**Table 2.** Equipment for gaseous emissions measurements

| HC | Horiba Instruments FMA-220 flame ionization analyzer |
|---|---|
| CO, CO$_2$ | Horiba Instruments AIA-220 infrared analyzer |
| NO$_x$ | Horiba Instruments CLA-220 chemiluminescent analyzer |

AMS FACILITY – The Center for AMS at LLNL houses four accelerators with different analysis capabilities [20-24]. The samples in this study were analyzed with the HVEE FN system operating at 6.5 MV (Fig. 2). All samples were prepared in the LLNL natural carbon prep lab using established methods [25]. The AMS sample prep method accommodates samples containing between 0.05 and 10 mg carbon. Samples containing 0.2-2 mg carbon are preferred for obtaining higher measurement precision and lower systemic backgrounds. Approximately 15000 $^{14}$C-AMS samples are measured annually at LLNL with 2-3 measurement days per week.

**Figure 2.** LLNL HVEE FN accelerator system viewed from the high energy end looking toward the ion source on the far side of the accelerator tank.

## MATERIALS AND METHODS

TEST FUELS - All major fuel components and lubrication oils were checked for $^{14}$C content prior to use in the engine. Diesel fuel was not separated into components, its isotopic content was measured with any additives or detergents added by the manufacturer. Baseline diesel from two production lots were used. The test fuels included baseline diesel fuel and blends of baseline diesel with various amounts of bio-derived ethanol. Because ethanol is soluble in diesel fuel in only small quantities, either an emulsifier (Span 85, i.e., sorbitan trioleate) or a cosolvent (n-butanol) was used to prepare the ethanol-in-diesel blends. An ignition improver, di-tert-butyl peroxide (DTBP), was also used to compensate for the low cetane number of ethanol and to give all blends the same cetane number (49.2) as the baseline diesel. Table 3 lists fuel blends and $^{14}$C content of the major constituents in the fuels. The $^{14}$C content of diesel fuel is higher than expected for a petroleum product. The addition of small amounts of contemporary carbon detergents are likely responsible for the elevation. The isotopic content of the lubrication oil was at instrumental background. The contemporary carbon from the ethanol is the tracer in these fuels. Since the

**Table 3**. UCB test fuel blends (components listed in volume percent except for oxygen which is mass percent) and ¹⁴C content of components (amol ¹⁴C/ mg C).

| Fuel Component | Blend 10E | Blend 10C | Blend 20E | Blend 20C | Blend 40E | Blend 40C | ¹⁴C Content |
|---|---|---|---|---|---|---|---|
| Diesel A | 88.0 | 88.5 | 76.0 | 77.0 | | | 0.33 |
| Diesel B | | | | | 52.0 | 54.0 | 0.23 |
| Ethanol | 9.5 | 9.0 | 18.5 | 18.0 | 37.0 | 36.0 | 107 |
| SPAN 85 | 2.0 | | 4.0 | - | 8.0 | | 110 |
| n-butanol | | 2.0 | - | 3.5 | | 7.0 | 0.09 |
| DTBP | 0.5 | 0.5 | 1.5 | 1.5 | 3.0 | 3.0 | 0.01 |
| Oxygen | 3.5 | 3.5 | 7.0 | 7.0 | 13.9 | 13.9 | NA |

ethanol is bio-derived, each carbon atom in the ethanol is equally labeled with ¹⁴C.

FILTER HANDLING AND LOADING – All PM samples were collected on 47 mm Gelman Sciences PALLFLEX tissuquartz 2500QAT-UP membrane filters. These quartz filters were pre-combusted at 1173 K for 3 h to remove all carbon residue and allowed to cool to 300 K in the furnace before removal. Blank filters were first conditioned to temperature and humidity overnight in petri dishes and then individually weighed with a Mettler UM 3 microbalance. Before sampling, the dilution ratio was adjusted to yield temperatures of the diluted exhaust below the required 325 K; resulting dilution ratios ranged between approximately 6 and 14. Diluted exhaust was drawn through each filter for 10 minutes and the filters were then removed, placed in petri dishes and once again conditioned overnight before weighing (see Figure 3). Three to five separate samples were taken for each point in the test matrix. A series blank was collected for each set of filters.

**Figure 3**. Filters loaded with PM from UCB engine. The filters from left to right are a blank control, a collection with 20% ethanol, and a collection with baseline diesel.

GAS SAMPLE COLLECTION AND PROCESSING – For each fuel used, three gas samples were collected in 3.0 L Tedlar bags placed after the filter holder in the exhaust line. Each bag had a conventional fill valve and second septa seal port. The septa port was used to remove gas for AMS sample preparation. Gas samples were processed within several days of collection. The bags were maintained at temperatures between 290 and 300 K and retained gas for weeks.

The diluted exhaust gases contained ~6% $CO_2$ and very low levels of CO. Approximately 40 mL of exhaust gas at atmospheric pressure was transferred to an evacuated stainless steel AMS graphitization line. The transfer volume was selected to produce an AMS sample containing approximately 1 mg carbon. Water was removed using a dry ice/ isopropanol cold trap. The $CO_2$ was then cryogenically condensed in a liquid nitrogen (LN) cold trap and non-condensable gases were removed. The $CO_2$ was then moved to a graphitization head [25] for conversion to an AMS graphite sample.

CO cannot easily be removed from gas by cryogenic methods because its condensation temperature is below that of LN, 78K. CO can be removed from a gas on 3A molecular sieve (MS) material (Sigma Chemical, St. Louis, MO). The MS was pre-combusted at 1173 K for 3-h to remove all residual carbon, allowed to cool to 300 K, and then stored in a tightly sealed glass bottle with a teflon lined top prior to use. Approximately 600 mg MS was used to capture a <0.1 mg CO sample. The same 40 mL volume used to produce $CO_2$ samples was used in preparing CO samples. Dry ice/alcohol and LN cold traps removed water, $CO_2$, HC, and $SO_2$ from the 40 mL volume. It was then allowed to contact MS in a quartz tube partially submersed in LN for several minutes. The residual gases were then removed and the process was repeated twice. The quartz tube containing the MS and about 20 mg copper oxide (CuO) was evacuated, sealed with a torch fueled with hydrogen and oxygen, and combusted like a typical AMS sample. After combustion, the sample was converted to graphite using the standard AMS procedure [25]. Typical CO samples contained 40-70 μg carbon. Process blanks prepared on the rig contained <10 μg carbon, too little to make graphite.

AMS FILTER PREPARATION – Beyond measuring total PM emissions, we sought to determine the susceptibility of PM to exhaust oxidation treatment to reduce mass. In practice, the criteria for this separation are operationally defined by the investigator. Depending on the field of the investigator, the separation of PM carbon is commonly described by the following pairs: elemental and organic carbon, soluble organic fraction (SOF) and insoluble organic fraction (IOF), or volatile organic fraction

(VOF) and non-volatile organic fraction (NVOF). We prefer the VOF/NVOF nomenclature because it reflects the process we use and the physical properties employed in any realistic exhaust treatment scheme.

Loaded filters are cut in half with a clean stainless steel surgical scissors. One half is cut into strips and placed in a quartz combustion tube with CuO oxidizer and converted to an AMS graphite sample [25]. This measurement is of the total carbon in the PM. The other half filter is heated to 613K for 2 h in a furnace and then allowed to cool to room temperature. This procedure to remove the VOF was developed using National Institute of Standards and Technology (NIST) standard reference material (SRM) to obtain consistent isotope ratios and mass fraction of the NVOF. NIST SRM 2975 (diesel soot) and SRM 1649a (urban dust) are the closest NIST SRMs to exhaust PM. The diesel soot SRM 2975 had only 7% VOF with this procedure, much less than observed with PM from a typical diesel. The filters loaded with PM lose mass during the thermal separation and the soot deposits are noticeably lighter. The remaining carbon is the NVOF. The filters with NVOF are then prepared as AMS samples with the usual procedure [25].

Gravimetric measurement of PM deposited on filters can be unreliable if the PM mass is small (100 µg) and the filter mass is large (100 mg). During the AMS sample preparation method we measure the $CO_2$ pressure from the completely combusted sample. The VOF is estimated by difference in mass of the two half filters.

AMS MEASUREMENT AND ANALYSIS - AMS is an isotope ratio mass spectrometry technique where $^{14}C/^{13}C$ ratios of the unknowns are normalized to measurements of 4-6 identically prepared standards of known isotope concentration. Typical samples are placed in quartz combustion tubes with excess copper oxide (CuO), evacuated and combusted to $CO_2$. The evolved $CO_2$ is purified, trapped, and converted to graphite in the presence of cobalt or iron catalyst in individual reactors [25]. Large $CO_2$ samples (> 500 µg) can be split for additional $^{13}C$ measurement by stable isotope ratio mass spectrometry. Identified fuel components were measured for $^{13}C$ and gave $\delta^{13}C$ corrections of −27 per 1000. All graphite targets were measured at the Center for AMS at LLNL.

The measured ratio of $^{14}C$ to total C for each sample, $R_{sample}$, is described in Eq. 1. The concentration of the $^{14}C$ labeled fuel component is $^{14}C_{tracer}/C_{tracer}$. The contributions from the fuel and additives to the measured ratio are $^{14}C_{fuel}/C_{fuel}$ and $^{14}C_{add}/C_{add}$, respectively. The background contribution is $^{14}C_{bk}/C_{bk}$ and the possibility of contamination to the sample is indicated as $^{14}C_{uk}/C_{uk}$.

$$R_{sample} = \frac{^{14}C_{tracer} + {}^{14}C_{fuel} + {}^{14}C_{add} + {}^{14}C_{bk} + {}^{14}C_{uk}}{C_{tracer} + C_{fuel} + C_{add} + C_{bk} + C_{uk}} \qquad (1)$$

In theory, all the components in Eq. 1 need to be determined by a series of control experiments. In practice some components can be minimized by experimental design. In the case of PM samples, the $^{14}C$ terms of petroleum derived fuel components are insignificant, only a biologically-derived additive (e.g., Span 85) contributes to the $^{14}C$ content. The $^{14}C_{bk}$ component is a systemic background of ambient $CO_2$ absorbed by the deposited PM. It is hoped that $C_{uk}$ is eliminated and $C_{bk}$ is consistently measured in blanks and baseline diesel samples.

The isotope ratio of the sample, $R_{sample}$, is calculated from the measured isotope ratios of the sample, $R_{sample(meas)}$, the average of the measured standards, $R_{stand(meas)}$, and the known isotope ratio of the standard, $R_{stand}$, shown in Eq. 2.

$$R_{sample} = \frac{R_{sample(meas)}}{R_{stand(meas)}} R_{stand} \qquad (2)$$

Traditional tracer experiments generally depend on radioactive decay and are dominated by a highly labeled tracer with very small mass. In our case, the tracer was not radioactive and contributed a significant amount of carbon to the measured isotope ratio. The denominator of the Eq. 1 is an expression for the total carbon mass of the sample. The product of the isotope ratio and carbon mass is the quantity of $^{14}C$ in the sample. The $^{14}C$ in the measured sample comes from the fuel components (see Eq. 3), whose isotope ratios we measure.

$$^{14}C_{sample} = {}^{14}C_{tracer} + {}^{14}C_{fuel} + {}^{14}C_{add} + {}^{14}C_{bk} \qquad (3)$$

The relative contribution of the tracer to the PM $^{14}C$ content then be calculated by solving Eq. 3 for the $^{14}C_{tracer}$ term. The carbon mass of the tracer, $C_{tracer}$, in the PM is determined by dividing the tracer $^{14}C$ content, $^{14}C_{tracer}$, by the tracer $^{14}C$ concentration, $R_{tracer}$, as in Eq. 4.

$$C_{tracer} = \frac{^{14}C_{tracer}}{R_{tracer}} \qquad (4)$$

The fraction of PM mass attributable to the tracer, $F_{tracer}$, is then the ratio of $C_{tracer}$ to sample carbon mass, $C_{sample}$, as shown in Eq. 5.

$$F_{tracer} = \frac{C_{tracer}}{C_{sample}} \qquad (5)$$

## RESULTS AND DISCUSSION

Brake-specific emissions and fuel consumption results from the baseline diesel fuel and the six test fuel blends are shown in Table 4. Since the different production lots of baseline diesel fuel used had different brake-specific emissions, comparison of the blends is also presented as a percentage of baseline diesel emissions in Table 5.

**Table 4.** Brake-specific emissions and fuel consumption of baseline diesel and ethanol-in-diesel blends in g/kW-hr.

| Fuel | Diesel A | Diesel B | Blend 10E | Blend 10C | Blend 20E | Blend 20C | Blend 40E | Blend 40C |
|---|---|---|---|---|---|---|---|---|
| PM | 0.038 | 0.0221 | 0.0323 | 0.0302 | 0.044 | 0.027 | 0.0184 | 0.0457 |
| $NO_x$ | 5.50 | 4.23 | 5.04 | 5.09 | 4.89 | 4.87 | 4.40 | 4.76 |
| HC | 0.097 | 0.114 | 0.132 | 0.147 | 0.155 | 0.155 | 0.291 | 0.310 |
| CO | 0.400 | 0.407 | 0.453 | 0.449 | 0.496 | 0.484 | 1.703 | 2.121 |
| Fuel Consumption | 185 | 184 | 201 | 207 | 206 | 208 | 227 | 225 |

**Table 5.** Brake-specific emissions and fuel consumption of ethanol-diesel blends expressed as percent of baseline diesel.

| Fuel | Blend 10E | Blend 10C | Blend 20E | Blend 20C | Blend 40E | Blend 40C |
|---|---|---|---|---|---|---|
| PM | 85% | 79% | 119% | 71% | 206% | 83% |
| $NO_x$ | 91% | 92% | 87% | 89% | 104% | 112% |
| HC | 136% | 151% | 155% | 160% | 255% | 271% |
| CO | 112% | 111% | 120% | 121% | 419% | 522% |
| Fuel Consumption | 109% | 112% | 111% | 112% | 124% | 123% |

The homogeneous cosolvent blends (Blends 10C, 20C, and 40C in Tables 4 and 5) reduced total gravimetric PM emissions in each case. The heterogeneous emulsified blends (Blends 10E, 20E and 40E) actually experienced an increases in total PM mass for the 20 and 40% ethanol blends. The loaded filters did not appear darker however, suggesting that additional mass may have been due to increased water or unburned fuel absorption on the PM. $NO_x$ emissions from the 10 and 20% blends were 90% that of the baseline diesel, probably due to lower combustion temperature. Emissions of both HC and CO increased for 10 and 20% blends, but remained very low as is typical with diesel engine combustion. Fuel consumption was higher with the test fuel blends due to the lower energy density of ethanol. The Cummins engine ran poorly with the 40% ethanol blends and emissions were unusually high for a diesel.

VOF/NVOF PARTITIONING – The $CO_2$ pressures measured from half filters during the production of graphite for AMS analysis provide an accurate measure of carbon mass. Traditional gravimetric measurements of filters can be misleading. It is difficult to measure 100 µg variations in a 100 mg filter. Furthermore, the filters are fragile and it is easy to lose small pieces on the soot-free border when handling with forceps. The PM deposited on quartz filters has very high specific surface area and readily absorbs water and other molecules from the vapor phase.

The partitioning of carbon in the VOF and NVOF for the different fuel blends is shown in Figure 4. The baseline diesel fuels, DA and DB, had NVOF/VOF splits of roughly 55/45 ± 1% and 48/52 ± 6%. The DB filters had significantly more variation. Partitioning of NVOF/VOF from the 10 and 20% ethanol cosolvent blends (10C and 20C) was essentially the same as the baseline fuels. The 10 and 20% ethanol emulsified blends (10E and 20E) had much higher VOF than baseline fuels, but samples were more variable (±10%). Both 40% ethanol blends produced low NVOF, but the engine ran poorly. Only the 10 or 20% ethanol blends are viable fleet fuels.

When viewed in conjunction with the brake-specific emissions, the large VOF of the 20% ethanol emulsified blend yields a brake-specific PM NVOF only 23% that of diesel assuming a 20-80 NVOF-VOF partition. The cosolvent blend yields a brake-specific PM NVOF 32% that of diesel assuming a 45-55 NVOF-VOF partition. The NVOF PM is more resistant to exhaust oxidation, so driving more PM into the VOF may be more effective in reducing emissions.

ETHANOL CONTRIBUTION TO PM – The [14]C signal from the bio-derived ethanol in the PM was easily measured above the isotope level of the baseline diesel. Blank filters used to measure the carbon mass and isotopic content of the ambient environment absorbed approximately 25 µg carbon with a [14]C content about 50% of contemporary plants. In a half filter this background accounted for 0.5 amol of [14]C. Filters

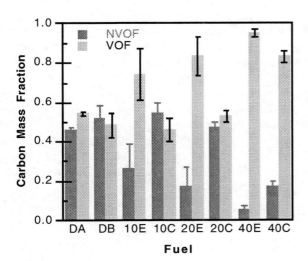

**Figure 4.** Partitioning of carbon between VOF and NVOF on filters loaded with PM from both baseline diesel fuels (DA and DB), emulsified blends containing 10, 20 and 40% ethanol (10E, 20E, and 40E) and cosolvent blends containing 10, 20 and 40% ethanol (10C, 20C, and 40C). Error bars represent standard deviation of 3-5 measurements.

loaded with PM from running baseline diesel absorbed [14]C between loading and AMS sample processing. Since fresh soot is more absorbent than aged soot, it is reasonable to assume most of this [14]C was absorbed in the engine lab where the ambient [14]C content is 50% that of contemporary plants. This absorbed mass is approximately 10 μg. Finally, the [14]C content of the diesel fuel is 0.3 amol / mg C. This level is higher than expected for a petroleum-derived material. Commercial diesel fuel contains a variety of additives, some of which may have bio-derived carbon, that slightly elevate the [14]C above a single petroleum derived molecule (e.g. n-butanol in this study). The PM collected from the baseline diesel fuels contained about 2.0 amol [14]C / mg C. The thermal treatment to remove the VOF cut this background level to 1.3 amol [14]C / mg C. The absorbed background depends on surface area which depends on the mass of PM deposited. Scaling the background as a function of sample mass is achieved by combining the fuel and background contributions and expressing them as a ratio of [14]C to PM sample mass.

The contributions of ethanol to the carbon in PM is listed in Table 6. The fraction of carbon mass contributed by ethanol varies slightly between emulsified and cosolvent blends so that the same oxygen content was achieved by each blending method. The ethanol carbon mass percentage was calculated from knowledge of the mixture components. The total and NVOF PM ethanol carbon mass percentage are averages of measurements from 2-5 samples. Uncertainties were driven by scatter among samples rather than measurement precision and typically ranged 0.2-0.7%.

**Table 6.** Distribution of ethanol-derived carbon in total and NVOF PM.

| Fuel Blend | Ethanol Vol % | Fuel Ethanol C Mass % | Total PM Ethanol C Mass % | NVOF PM Ethanol C Mass % |
|---|---|---|---|---|
| 10E | 9.5 | 5.7 | 2.4 | 3.4 |
| 10C | 9.0 | 5.4 | 0.8 | 2.7 |
| 20E | 18.5 | 11.7 | 3.7 | 10.8 |
| 20C | 18.0 | 11.4 | 2.9 | 6.7 |
| 40E | 37.0 | 26.2 | 20.9 | 24.2 |
| 40C | 36.0 | 25.2 | 3.7 | 15.0 |

The following trends were observed during the study of blended fuels:

- Carbon masses of the PM generated by the cosolvent and emulsified blends were almost identical.
- PM from the cosolvent blend had a lower [14]C concentration than that of the emulsified blend.
- Ethanol-derived carbon residing in PM is more likely to be NVOF.
- The [14]C concentrations of the PM were below those of the $CO_2$ for the 10% and 20% blends.

The trend that ethanol-derived carbon was less likely to be found in the PM is not surprising. Carbon mass measured by $CO_2$ pressure from combusted half filters was the same for cosolvent and emulsified blends. The emulsified blend produced PM with higher mass but did not contain more carbon. Furthermore, much of this larger mass was volatile. The combination of no carbon and volatility of the PM and the possibility of unburned emulsifier in the PM suggest that the higher PM mass is due to absorbed water. In the case of the cosolvent blends, it appears that the ethanol-derived carbon resides primarily in the building blocks of soot rather than in condensed volatiles on the surface of the PM. Ethanol produces some acetylene and other soot precursors during combustion which contribute to soot formation. Although its contribution to soot is less than diesel fuel, the oxygenate does participate in soot formation. Some of the loss of [14]C to the VOF for the emulsified blend appears to be associated with the emulsifier. Control experiments in which emulsifier was added to diesel without ethanol indicated that emulsifier-derived carbon resided mostly in the VOF.

AMS GAS ANALYSIS – Since the majority of the carbon from the fuel is fully combusted, the isotope ratio of the collected $CO_2$ reflects the isotopic content of the fuel. The $CO_2$ from the emulsified blend has more [14]C than the cosolvent blend as expected due to use of a contemporary emulsifier. The [14]C content of fuels and collected $CO_2$ is listed in Table 7. The collected $CO_2$ samples had more [14]C than the fuel average in all cases. A small elevation is due to the small amount of atmospheric $CO_2$ drawn through the engine and possible migration through the collection bag. Additionally, the ethanol in the blends is more likely to reach $CO_2$ than the diesel fuel, slightly increasing the measured [14]C content of the combusted gas.

The CO samples were so small that high precision measurement of [14]C content was impossible. Some CO samples had slight isotopic elevation compared to $CO_2$, indicating that the ethanol may not have combusted as fully as the diesel fuel. In general, however, isotopic measurements of separated CO were the same as the $CO_2$ within measurement uncertainties.

**Table 7**. Comparison of fuel and combusted gas $CO_2$ [14]C content for baseline diesel A and fuel blends.

| Fuel | Fuel (amol [14]C / mg C) | $CO_2$ (amol [14]C / mg C) |
|---|---|---|
| Diesel | 0.33 | 0.86 |
| 10E | 8.10 | 8.63 |
| 10C | 6.12 | 6.55 |
| 20E | 16.3 | 17.4 |
| 20C | 12.5 | 13.5 |
| 40E | 35.9 | 37.7 |
| 40C | 27.2 | 28.1 |

## CONCLUSION

AMS provides a means of following the fate of carbon in specific compounds from the fuel to the emissions from diesel engines. Selective labeling of specific carbon atoms within a fuel component provides direct experimental evidence of the behavior of different chemical groups during combustion. In addition to providing data validation to combustion modelers, the data provides insights into which chemical structures within fuels and additives most greatly influence emissions.

Solubility issues and extremely low cetane number limit the likelihood that ethanol would be used routinely as a diesel oxygenate. The ease of using the contemporary tracer carbon demonstrated the simple application to tracing bio-derived components of blended fuels. Despite being partially oxidized, ethanol-derived carbon contributed significantly to PM. The relatively large amounts of ethanol-derived carbon in the PM points to potential problems of using oxygenates which form unsaturated C2 fragments during combustion. The heterogeneous emulsified blends had larger VOF than the homogeneous cosolvent blends. The differences in the PM produced with cosolvent and emulsified blends

indicates that the distribution of oxygen in the fuel, not just its content, significantly affects PM production.

## ACKNOWLEDGMENTS

This work was performed under the auspices of the U.S. Department of Energy by University of California Lawrence Livermore National Laboratory under Contract No. W-7405-Eng-48. The engine experiments reported were conducted at The Combustion Laboratories of the Department of Mechanical Engineering at the University of California at Berkeley. The research at UC Berkeley was supported by LLNL Laboratory Directed Research and Development grant 01-ERI-007.

## REFERENCES

1. Cheng, A. S. and R. W. Dibble. "Emissions Performance of Oxygenate-in-Diesel Blends and Fischer-Tropsch Diesel in a Compression Ignition Engine," SAE paper 1999-01-3606.
2. Wong, G., B. L. Edgar, T. J. Landheim, L. P. Amlie, R. W. Dibble and D. B. Makel. "Low Soot Emission from a Diesel Engine Fueled with Dimethyl and Diethyl Ether," WSS/CI paper 95F-162, October 1995.
3. Liotta, F. J. and D. M. Montalvo. "The Effect of Oxygenated Fuels on Emissions from a Modern Heavy-Duty Diesel Engine," SAE paper 932734.
4. Fleisch, T, C. McCarthy, A. Basu, C. Udovich, P. Charbonneau, W. Slodowske, S. Mikkelsen. And J. McCandless. "A New Clean Diesel Technology: Demonstration of ULEV Emissions on a Navistar Diesel Engine Fueled with Dimethyl Ether," SAE paper 950061.
5. McCormick, R. L., J. D. Ross and M. S. Graboski. "Effect of Several Oxygenates on Regulated Emissions from Heavy-Duty Diesel Engines," *Environmental Science & Technology*, **31** 144-1150, 1997.
6. Miyamoto, N., H. Ogawa, N. M. Nurun, K. Obata and T. Arima. "Smokeless, Low $NO_x$, High Thermal Efficiency, and Low Noise Diesel Combustion with Oxygenated Agents as Main Fuel," SAE paper 980506.
7. Maricq, M. M., R. E. Chase, D. H. Podsiadilik, W. O. Siegel and E. W. Kaiser. "The Effect of Dimethoxy Methane Additive on Diesel Vehicle Particulate Emissions," SAE paper 982572.
8. Bertoli, C., N. Del Giacomo and C. Beatrice. "Diesel Combustion Improvements by the Use of Oxygenated Synthetic Fuels," SAE paper 972972.
9. Cheng, A. S., R. W. Dibble, and B. A. Buchholz. "Isotopic Tracing of Particulate Matter from a Compression-Ignition Engine Fueled with Ethanol-in-Diesel Blends," Preprint from the Symposium on Chemistry of Liquid and Gaseous Fuels, Division of Fuel Chemistry, 219th American Chemical Society Meeting, San Francisco, March 26-30, 2000. pp.288-93.

10. Wang, W. G., D. W. Lyons, N. N. Clark, M. Gautam, P. M. Norton. "Emissions from Nine Heavy Duty Trucks by diesel and Biodiesel blend without Engine Modification," *Environ. Sci Tech.* **34**: 933-939, 2000.

11. Gonzalez, M. A., W. J. Piel, T. W. Asmus, W. Clark, J. A. Garbak, E. Liney, M. Natarajan, D. W. Naegeli, D. Yost, E. A. Frame and J. P. Wallace III. "Screening for Advanced Petroleum-Based Diesel Fuels: Part 2—The Effect of Oxygenate Blending Compounds on Exhaust Emissions," SAE paper 2001-01-3632.

12. Cheng, A. S. and R. W. Dibble. "Emissions from a Cummins B5.9 Diesel Engine Fueled with Oxygenate-in-Diesel Blends," SAE paper 2001-01-2505.

13. Kass, M.D., J.F. Thomas, J. M. Storey, N. Domingo, J. Wade and G. Kenreck. "Emissions from a 5.9 Liter Diesel Engine Fueled with Ethanol Diesel Blends," SAE Paper 2001-01-2018.

14. Cole, R.L., R.B. Poola, R. Sekar, J.E. Schaus and P. McPartin. "Effects of Ethanol Additives on Diesel Particulate and $NO_x$ Emissions," SAE Paper 2001-01-1937.

15. Homan, H. S. and W. K. Robbins. "A Carbon-14 Tracer Study of the Relative Fractions of Various Fuel Carbons in Soot," *Combustion and Flame* **63**: 177-190, 1986.

16. Schmeider, R. W. "Radiotracer Studies of Soot Formation in Diffusion Flames," Twentieth International Symposium on Combustion, The Combustion Institute, 1025-1033, 1984.

17. Vogel, J.S., K. W. Turteltaub, R. Finkel and D. E. Nelson. "Accelerator Mass Spectrometry – Isotope Quantification at Attomole Sensitivity," *Anal. Chem.* **67**: A353-A359, 1995.

18. Ball, J., C. A . Chappin, J. P. Buckingham, E. A. Frame, D. M. Yost, M. Gonzalez, E. M. Liney, M. Ntarajan, J. P. Wallace and J. A. Garbak. "Dimethoxymethane in Diesel Fuel: Part 1 – The Effect of Operating Mode and Fuels on Emissions of Toxic Air Pollutants and Gas and Solid Phase PAH," SAE paper 2001-01-3627.

19. Ball, J., C. A . Chappin, J. P. Buckingham, E. A. Frame, D. M. Yost, M. Gonzalez, E. M. Liney, M. Ntarajan, J. P. Wallace and J. A. Garbak. "Dimethoxymethane in Diesel Fuel: Part 2 – The Effect of Fuels on Emissions of Toxic Air Pollutants and Gas and Solid Phase PAH Using a Composite of Engine Operating Modes," SAE paper 2001-01-3628.

20. Roberts M.L., G. S. Bench, T. A. Brown, M. W. Caffee, R. C. Finkel, S. P H. T. Freeman, L. J. Hainsworth, M. Kashgarian, J. E. McAninch, I. D. Proctor, J. R. Southon and J. S. Vogel. "The LLNL AMS Facility," *Nucl. Instru. Meth. B,* **123**: 57-61, 1997.

21. Southon, J. and M. Roberts. "Ten Years of Ion Sourcery at CAMS/LLNL - Evolution of a Cs ion source," *Nucl. Instru. Meth. B,* **172**: 257-261, 2000.

22. Roberts, M. L. , R. W. Hamm, K. H. Dingley, M. L. Chiarappa-Zucca and A. H. Love. "A Compact Tritium AMS System," *Nucl. Instru. Meth. B,* **172**: 262-267, 2000.

23. T. J. Ognibene, T. A. Brown, J. P. Knezovich, M. L. Roberts, J. R. Southon and J. S. Vogel. " Ion-optics calculations of the LLNL AMS system for biochemical C-14 measurements," *Nucl. Instru. Meth. B,* **172**: 47-51, 2000.

24. Roberts, M.L., P. G. Grant, G. S. Bench, T. A. Brown, B. R. Frantz, D. H. Morse and A. J. Antolak. "The stand-alone microprobe at Livermore," *Nucl. Instru. Meth. B,* **158**: 24-30, 1999.

25. Vogel, J.S., J. R. Southon and D. E. Nelson. "Catalyst and Binder Effects in the Use of Filamentous Graphite for AMS," *Nucl. Instrum. Methods Phys. Res. Sect. B* **29**: 50-56, 1987.

## CONTACT

Bruce A. Buchholz
CAMS, L-397
Lawrence Livermore National Laboratory
P.O. Box 808
Livermore, CA 94551-9900
VOI: 925-422-1739
FAX: 925-423-7884
Email: buchholz2@llnl.gov

## DEFINITIONS, ACRONYMS, ABBREVIATIONS

$^{14}C$: carbon-14, long-lived naturally occurring radioisotope of carbon

**AMS**: accelerator mass spectrometry

**amol**: attomole, $1 \times 10^{-18}$ mole, $\sim 6 \times 10^5$ atoms

**DTBP**: ditertiarybutyl peroxide

**HC**: hydrocarbons

**LLNL**: Lawrence Livermore National Laboratory

**LN**: liquid nitrogen

**MS**: molecular sieve

**NVOF**: non-volatile organic fraction

**PM**: particulate matter

**UCB**: University of California at Berkeley

**VOF**: volatile organic fraction

2002-01-1705

# The Effect of Oxygenates on Diesel Engine Particulate Matter

**A. S. (Ed) Cheng and Robert W. Dibble**
University of California, Berkeley

**Bruce A. Buchholz**
Lawrence Livermore National Laboratory

## ABSTRACT

A summary is presented of experimental results obtained from a Cummins B5.9 175 hp, direct-injected diesel engine fueled with oxygenated diesel blends. The oxygenates tested were dimethoxy methane (DMM), diethyl ether, a blend of monoglyme and diglyme, and ethanol. The experimental results show that particulate matter (PM) reduction is controlled largely by the oxygen content of the blend fuel. For the fuels tested, the effect of chemical structure was observed to be small. Isotopic tracer tests with ethanol blends reveal that carbon from ethanol does contribute to soot formation, but is about 50% less likely to form soot when compared to carbon from the diesel portion of the fuel.

Numerical modeling was carried out to investigate the effect of oxygenate addition on soot formation. This effort was conducted using a chemical kinetic mechanism incorporating n-heptane, DMM and ethanol chemistry, along with reactions describing soot formation. Results show that oxygenates reduce the production of soot precursors (and therefore soot and PM) through several key mechanisms. The first is due to the natural shift in pyrolysis and decomposition products. In addition, high radical concentrations produced by oxygenate addition promote carbon oxidation to CO and $CO_2$, limiting carbon availability for soot precursor formation. Additionally, high radical concentrations (primarily OH) serve to limit aromatic ring growth and soot particle inception.

## INTRODUCTION

Previous studies have shown that diesel engines operating on neat oxygenated fuels produce virtually no particulate matter (PM) emissions [1-6]. However, many oxygenates have fuel properties that make them unattractive as a diesel fuel and some (e.g., dimethyl ether (DME)) require highly modified or even redesigned fuel delivery systems.

These disadvantages can be overcome by using oxygenates as a blending agent for conventional diesel fuel. Oxygenated diesel blends have been shown to dramatically reduce PM emissions from diesel engines, to an extent (% reduction) far greater than their amount of addition (% of fuel by volume or mass) [7-26]. The PM reductions are achieved with little or no change to $NO_x$ emissions, resulting in a favorable shift in the $NO_x$-PM tradeoff curve. With a fuel-induced reduction in PM, engine modifications can be subsequently employed to reduce $NO_x$, with the overall effect being a simultaneous reduction in both pollutants. This is illustrated conceptually in Figure 1.

Beyond the emissions benefits, oxygenated diesel blends can help to reduce foreign oil dependence and

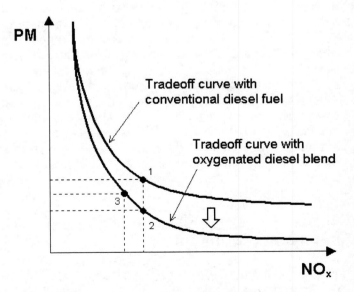

**Figure 1.** Oxygenated diesel blends produce a shift in the $NO_x$-PM tradeoff curve. The effect of the fuel change might result a movement from Point 1 to Point 2 in the figure. Subsequent use of engine modifications for $NO_x$ control could then move emissions along the new tradeoff curve to a level represented by Point 3.

promote the use of renewable energy sources. A number of oxygenates can be produced from non-petroleum sources. For example, dimethoxy methane (DMM) and DME can be manufactured from gas-to-liquids processes using natural gas as the feedstock [27]. Glycol ethers such as monoglyme and diglyme can be derived using a syngas process with coal as the feedstock [28]. Certain oxygenates can also be developed as renewable fuels. The most common is ethanol, which can be produced from corn or other biomass. Such bio-derived ethanol can be further manufactured into other oxygenates, such as diethyl ether (DEE).

While the benefits of oxygenated diesel blends are evident, a thorough understanding of the mechanisms that bring about the reductions in PM are not. Many researchers have indicated that fuel oxygen content is the main factor affecting PM emissions. For example, the results of Miyamoto et al. are often cited which show a decrease in Bosch smoke number that is well correlated to fuel oxygen content, with smoke levels becoming essentially zero at an oxygen content of approximately 30% by weight (mass) [13]. However, others have concluded that there are important differences depending on the chemical structure or volatility of a given oxygenate [7, 11, 20, 22]. To further investigate the mechanisms, a number of investigators have carried out numerical modeling of the chemical kinetics in the primary soot formation region [29-33]. These studies provide additional insight into the nature of PM reduction with oxygenated diesel blends.

The current paper has two objectives. The first is to present a summary of experimental results obtained for a variety of oxygenated blends using a Cummins B5.9 175 hp, direct-injected (DI) diesel engine. A number of conclusions are drawn based on the aggregate experimental data. The second objective is to carry out numerical simulations of the effect of oxygenates on soot precursor formation and soot particle inception. This effort was conducted using a chemical kinetic mechanism incorporating n-heptane, DMM and ethanol chemistry, along with reactions describing soot formation. The ultimate goal is to identify the factors that govern PM reductions, so that specific criteria can be used to select the most suitable oxygenated blend fuels for diesel engines.

## SUMMARY AND DISCUSSION OF PREVIOUS EXPERIMENTAL RESULTS

The authors have previously presented experimental results from oxygenated blend tests using DMM, DEE, a blend of monoglyme and diglyme called Cetaner, and ethanol [24-26]. The chemical structures and selected properties of these fuels are shown in Table 1. The engine tests were conducted on a 1993 Cummins B5.9 175 hp, 6-cylinder, turbocharged and aftercooled, DI

**Table 1.** Chemical structure and selected properties of oxygenated fuels tested.

| Fuel component | Chemical Formula | Oxygen content (mass%) | Boiling point (°C) | Heating value (MJ/L) |
|---|---|---|---|---|
| Dimethoxy methane (DMM) | $CH_3OCH_2OCH_3$ | 42.1 | 42 | 20.2 |
| Diethyl ether (DEE) | $CH_3CH_2OCH_2CH_3$ | 21.6 | 34 | 24.1 |
| Cetaner = 20% monoglyme + | $CH_3O(CH_2)_2OCH_3$ | 35.5 | 85 | 25.2 |
| 80% diglyme | $CH_3O(CH_2)_2O(CH_2)_2OCH_3$ | 35.8 | 162 | 26.6 |
| Ethanol | $C_2H_6OH$ | 34.7 | 78 | 21.2 |
| Diesel | various | — | 173-360 | 35.4 |

diesel engine. Fuel injection was mechanically governed, although the injection pump was capable of relatively high injection pressures of up to 115 MPa.

Figure 2 presents a summary of the PM reductions achieved with all of the oxygenated blends tested, relative to the baseline diesel fuels. The figure shows modal-averaged data for steady-state tests run with each of the fuels. The reader is referred to references 24-26 for a detailed description of the experimental procedures and results.

Within the experimental uncertainty, the data does suggest that the oxygen content of a given fuel blend is the predominant factor affecting its ability to reduce PM emissions. Changes in thermophysical properties of the test fuels (cetane number, volatility, energy density) certainly would have affected fuel injection and vaporization, ignition delay times, and heat release rates. Oxygenates with lower oxygen content (within the oxygenate itself) also required higher concentrations to produce a given oxygen level in the blended fuel. This served to displace more of the aromatics and sulfur contained in the baseline fuel and enhance the amount of PM reduction. However, all of these effects appear to

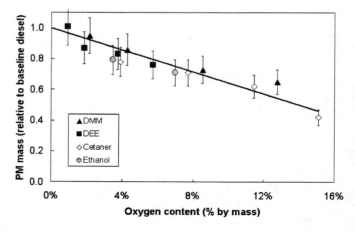

**Figure 2.** Experimental results of relative PM mass versus oxygen content for four types of oxygenated blend fuels.

have a smaller impact on PM than the blended fuel's oxygen content.

In addition, for the fuels tested, the effect of chemical structure was observed to be small. The oxygenates DEE, Cetaner, and ethanol contain C-C bonds, while DMM does not. The suggestion has been made that an oxygenate that lacks any C-C bonds is less likely to contribute to soot formation and PM mass as it cannot readily form important soot precursors such as acetylene ($C_2H_2$). However, the experimental data does not support this theory; DMM in fact appeared to be slightly less effective at reducing PM than the other oxygenates. It should be noted however that oxygenates with a wide variety of chemical structures were not tested. The three ethers and the one alcohol had at most two carbon atoms bonded together. Experimental tests conducted by Hallgren and Heywood showed that oxygenates possessing more complex or partial ring structures (diethyl maleate ($C_8H_{12}O_4$) and propylene glycol monomethyl ether acetate ($C_6H_{12}O_3$)) are much less effective for PM reduction when compared to diglyme [20].

Linear extrapolation of a best-fit line to the aggregate data indicates that oxygenate addition would reduce PM emissions to essentially zero at a oxygen content of 28%. This is in good agreement with the results of Miyamoto et al., although it cannot be stated with certainty that a linear relationship exists beyond the oxygenate levels tested.

The experiments conducted with DMM, DEE and Cetaner were carried out at numerous steady-state test modes (8 or 9 engine speed-load conditions) [24, 25]. Individual modal results revealed that oxygenate addition was more beneficial at high load conditions. At lower load conditions, PM emissions were not significantly reduced and in some cases even increased (relative to the baseline diesel fuel). Two factors likely contributed to this effect. The first is that less fuel is injected at lower power modes and therefore less fuel is burned during the mixing-controlled phase of combustion. Since soot formation occurs primarily during this mixing-controlled phase, the effect of the oxygenates on PM would be less pronounced at these engine modes. In addition, the overall (absolute) level of PM at the lower power modes was small. The oxygenated blends produced higher hydrocarbon (HC) emissions, and the contribution to PM mass from adsorbed or condensed HCs may have countered or overwhelmed a small reduction in inorganic PM mass.

An important observation was obtained from the tests with the ethanol blended fuels [26]. Because oxygenates do not produce PM when used as a neat fuel, it has been theorized that the oxygenates themselves do not participate in soot formation during the combustion of a oxygenated blend. For the ethanol blend experiments, accelerator mass spectrometry

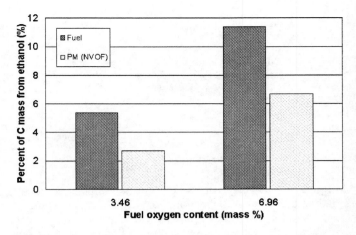

**Figure 3.** Percent of carbon mass from ethanol in the test fuels and in the non-volatile organic fraction (NVOF) of collected PM.

(AMS) was used to trace the carbon from grain ethanol, which possesses highly elevated carbon-14 radioisotope ($^{14}C$) levels compared to the petroleum-derived diesel fuel. Results from these tests are shown in Figure 3. As the figure indicates, ethanol carbon does participate in the formation of soot, but is about 50% less likely to form soot when compared to carbon originating from the diesel portion of the fuel.

## NUMERICAL MODELING OF SOOT FORMATION

The engine experiments discussed in the previous section provide valuable insight into the nature of PM reduction with oxygenated blend fuels. The results highlight the importance of fuel oxygen content and indicate that combustion chemistry is the major factor governing the ability of oxygenates to reduce diesel PM emissions. To further investigate this chemical effect, it is desirable to conduct numerical modeling of oxygenated fuel combustion under diesel-like conditions.

A comprehensive modeling effort would incorporate a detailed chemical kinetic mechanism for the oxygenated blend fuels along with a complete fluid dynamic description of the combustion chamber environment. However, this is infeasible from a practical standpoint due to current limits in computational power. Because the focus was to be on the chemical aspects of oxygenated fuel combustion, a strategy was employed to investigate detailed chemistry interactions while simplifying the fluid dynamics of the problem. Such a strategy can be justified based on recent insights obtained on the diesel combustion process by Dec and Flynn et al. [29, 33]. Their laser-sheet imaging work suggests that soot formation occurs primarily during the mixing-controlled phase of combustion, in a region between a standing fuel-rich premixed flame and the burning fuel jet's outer diffusion flame (Figure 4). This can be further conceptualized by following a parcel of injected fuel as shown in Figure 5. Fuel initially mixes with air, then is partially consumed in the rich premixed

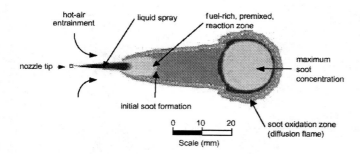

**Figure 4.** Conceptual view of diesel engine combustion during the mixing-controlled phase (the phase where the majority of soot formation is believed to take place). Soot is formed in the region between the fuel-rich, premixed flame and the outer diffusion flame [33].

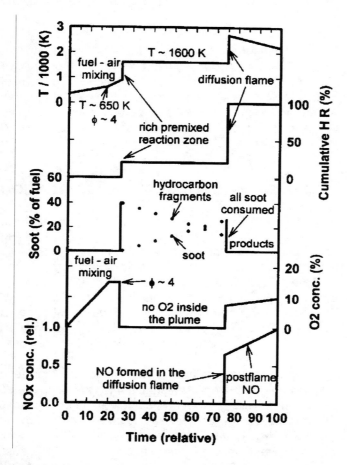

**Figure 5.** Depiction of processes that take place as a parcel of fuel is injected, becomes partially consumed in the premixed reaction zone, and then encounters the outer diffusion flame sheath [29].

reaction zone, where the fuel-air equivalence ratio is approximately $\phi \cong 4.0$. Soot formation occurs beyond this zone in a region where the products of rich combustion lead to polycyclic aromatic hydrocarbon (PAH) growth and particle inception. All or most of the soot formed is then oxidized upon encountering the diffusion flame sheath, where OH concentrations are high.

The current modeling effort therefore centers on the soot formation region inside the jet (the region between about 25 and 75 on the relative time scale of Figure 5). This region can be represented by a perfectly-mixed (0-D), constant pressure reactor. Flynn et al. [29] and Curran et al. [30, 31] have reported on changes to $C_2H_2$, ethylene ($C_2H_4$), and propargyl ($C_3H_3$) concentrations in this region due to the addition of various oxygenates. More recently, Kitamura et al. [32] has predicted the effect of oxygenates on PAH formation. The current effort utilizes a mechanism that carries the soot chemistry all the way to initial particle inception. The effect of oxygen levels is investigated using two oxygenates: DMM and ethanol. The objective is to study the evolution of soot concentrations within the soot formation region and also to predict soot levels at the location of the diffusion flame.

MECHANISM DEVELOPMENT

The overall reaction mechanism was developed beginning with a mechanism for n-heptane ($C_7H_{16}$) that included a chemical description of soot formation and oxidation. Although diesel fuel is actually comprised of many different hydrocarbon compounds, n-heptane serves as a good chemical surrogate for diesel and has cetane number similar to that of typical diesel fuels (CN $\cong$ 56) [34].

Reaction chemistry for DMM and ethanol was added to the n-heptane + soot mechanism using data from two other individual mechanisms. The combined reaction mechanism used in the numerical calculations of this paper consists of 159 species and 936 reactions.

N-Heptane + Soot Mechanism

The n-heptane + soot mechanism used was developed by Golovitchev et al. [34, 35] and is derived from the models of Westbrook et al. (n-heptane) [36-41] and Frenklach and Wang (soot) [42-44]. These models have been described extensively in the literature, but a summary of the key processes will be given here.

For the n-heptane chemistry, at the high-temperature conditions of diesel combustion (T > 900 K), initiation steps consist of unimolecular n-heptane decomposition as well as H-atom abstraction. The overall reaction then proceeds primarily through decomposition ($\beta$-scission) of the alkyl radicals, although $O_2$ addition can play an important role at more moderate temperatures (T $\cong$ 900 K).

The smaller gas-phase products of the n-heptane oxidation then serve as the building blocks for aromatic formation and PAH growth. The soot model begins with a series of reactions involving $C_2H_2$ and molecular hydrogen which lead to the formation of the phenyl radical ($A_1^\cdot$). This pathway is shown in Figure 6a.

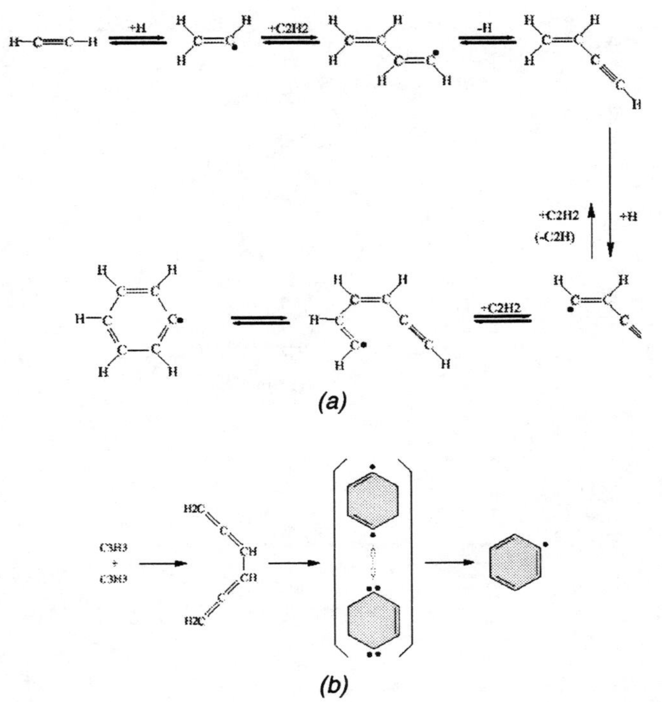

(a)

(b)

**Figure 6.** Formation of the phenyl radical from *(a)* reactions involving acetylene ($C_2H_2$) and *(b)* combination of two propargyl radicals ($C_3H_3$) [34].

Alternatively, the phenyl radical can be formed from two propargyl radicals ($C_3H_3$) as shown in Figure 6*b*. Following the formation of the first aromatic ring ($A_1$ = benzene), reactions involving H-atom abstraction and $C_2H_2$ addition (the so-called HACA mechanism) lead to aromatic ring growth. This mechanism is as follows:

$$A_i + H \rightarrow A_i^{\cdot} + H_2 \qquad (1)$$
$$A_i^{\cdot} + C_2H_2 \rightarrow A_iC_2H_2 \qquad (2)$$
$$A_iC_2H_2 + H \rightarrow A_iC_2H + H_2 \qquad (3)$$
$$A_iC_2H + C_2H_2 \rightarrow A_{i+1} \qquad (4)$$

where i represents the number of aromatic rings in a given species. The growth aromatic rings is included up to four-ring species. At that point, a prompt transition to soot (soot inception) is assumed to occur, i.e.,

$$A_4 \rightarrow 16C(s) + 5H_2 \qquad (5)$$

To limit aromatic ring growth, oxidation by OH and $O_2$ is included, using rate parameters from Frenklach and Wang [44].

## DMM Mechanism

Chemistry for DMM ($CH_3OCH_2OCH_3$) combustion was considered in the overall reaction mechanism based on the model of Naegeli et al. [45]. Initial reactions of the parent fuel are those of unimolecular decomposition and H-atom abstraction. Additional reactions with radical species produce dimethyl ether ($CH_3OCH_3$) and methyl formate ($CH_3OCHO$), which further react to yield radical

and smaller species. The methoxy methyl radical ($CH_3OCH_2$) is believed to play a particularly important role in the DMM kinetics through its reactions with oxygen ($O_2$) [45]. Also included in the mechanism are reactions that describe formaldehyde ($CH_2O$) chemistry.

The DMM mechanism was combined with the n-heptane + soot mechanism described above to produce a combined chemical kinetic model that could describe the behavior of DMM blends. In cases where there were duplicate reactions with different rate constants, the reactions from the n-heptane + soot mechanism were retained. This was because the n-heptane chemistry was more detailed and therefore involved a greater number of interdependent reactions that would be impacted by any changes in rate constants. Also, since the soot formation model was developed in conjunction with the n-heptane chemistry, it was desirable to retain as much of this original combination as possible.

## Ethanol Mechanism

The chemical kinetics of ethanol ($C_2H_5OH$) were included in the overall mechanism using reaction data developed by Marinov [46]. Of particular importance in the ethanol chemistry are the reactions that decompose $C_2H_5OH$ to $CH_3 + CH_2OH$, $C_2H_5 + OH$, $CH_3 + CH_2O$, and $CH_3HCO + H$. The evolution of intermediate and product species are also strongly influenced by reactions describing OH attack on the parent fuel. As with the DMM chemistry, ethanol species and reactions were added to the combined reaction mechanism (with the exception of those which would lead to duplication).

## Mechanism Validation

Both the n-heptane + soot and ethanol mechanisms have been extensively tested against experimental data from flow reactors, jet-stirred reactors, shock tubes, and rapid compression machines [34, 37, 42, 45, 46]. It was uncertain, however, whether the chemical kinetics described by the original mechanisms would be significantly altered by combining the mechanisms together. To investigate this possibility, model runs were conducted to compare ignition delay times predicted by the combined mechanism to those determined with the original mechanisms. The ignition delay times were computed using a constant volume, adiabatic, well-mixed (0-D) computational code [47].

Results for n-heptane – air mixtures are shown in Figure 7. Figure 7*a* shows the original data from the Golovitchev mechanism, while Figure 7*b* presents results form the current combined model. As the figures indicate, agreement between the two cases is very good. The combined mechanism properly exhibits the negative temperature coefficient (NTC) behavior in the temperature region that is associated with diesel combustion. There are slight differences in predicted ignition delays at lower temperature, fuel-lean conditions

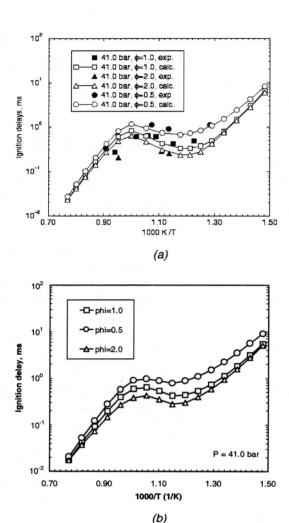

*(a)*

*(b)*

**Figure 7.** Mechanism validation for n-heptane – air mixtures. *(a)* Original Golovitchev mechanism results (open symbols) versus experimental data (solid symbols) [34], and *(b)* numerical results using combined mechanism.

(T < 800 K, $\phi$ = 0.5). However, this would have little impact on the current investigation which models reactions occurring under higher temperature, fuel-rich conditions.

Figure 8 shows comparisons between ignition delay results from the Marinov ethanol mechanism (Figure 8*a*) and results obtained from the combined mechanism (Figure 8*b*). The reactants are a mixture of 2.5% ethanol, 7.5% $O_2$ and 90% argon (mole fractions). Again, the combined mechanism shows excellent agreement with the original numerical results.

In contrast to the situation for the n-heptane and ethanol mechanisms, the chemical kinetics of the DMM mechanism cannot been rigorously tested due to the lack of experimental data available on the fundamental combustion behavior this fuel. The evolution of pyrolysis products has been measured by Edgar et al. [45] and those results do compare favorably with the numerical results of both the original DMM mechanism and the combined reaction mechanism. However, since

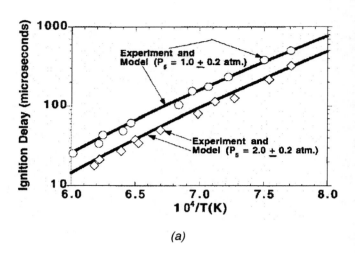

*(a)*

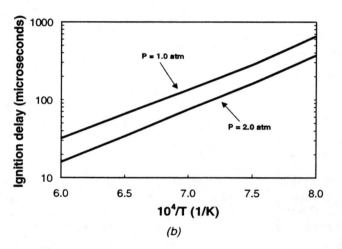

*(b)*

**Figure 8.** Mechanism validation for a mixture of 2.5% ethanol, 7.5% $O_2$, and 90% argon. *(a)* Comparison between Marinov mechanism and experimental results [46], and *(b)* numerical results using combined mechanism.

the mechanism has not been tested against a large number of experimental measurements, the validity of the DMM reaction kinetics cannot be determined with absolute certainty.

COMPUTATIONAL CODE

As previously mentioned, the numerical modeling effort focuses on the fuel-rich soot formation region that can be represented by a homogeneous, constant pressure reactor. Model runs were carried out assuming adiabatic, well-mixed, constant pressure conditions, similar to the strategy employed by Flynn et al. and Curran et al. [29-31]. The computational code utilizes the Chemkin-II interpreter and subroutine package to carry out the chemical kinetic computations and solve the governing conservation equations (Chemkin Interpreter version 3.6 and Chemkin Subroutine Library version 4.9). A full description of the Chemkin software can be found in References 48 and 49.

For all of the computations, an initial temperature of $T_i = 1000$ K and an initial pressure of $P_i = 10$ MPa were used. These values are representative of combustion conditions in the Cummins B5.9 engine, as estimated based upon pressure transducer data [50] and analogy to the conditions measured by laser diagnostics in Dec's experimental engine [33]. The initial fuel-air equivalence ratio for all cases was set at $\phi = 4.0$, which matches the approximate conditions of the standing premixed flame [29, 33]. During diesel engine combustion, the addition of oxygenates might alter to some extent the equivalence ratio in the mixture of injected fuel and entrained hot air (at a given point from the injector). However, since the standing premixed flame would develop at a location determined in large part by the equivalence ratio (for a given engine), it is fairly reasonable to maintain a constant value for $\phi$ regardless of the fuel oxygen content. Also, from a modeling standpoint, it was desirable to separate the effect of oxygenated fuel chemistry from any effects arising simply due to differences in equivalence ratio.

## RESULTS AND DISCUSSION

The analysis of numerical results focuses on soot concentrations and the precursor species $C_2H_2$, $C_3H_3$, and the first aromatic ring $A_1$. The evolution of these species inside the jet plume varies depending on the particular species under consideration. $C_3H_3$ is created almost exclusively in the fuel-rich premixed flame, where it experiences a sharp peak in concentration and is then rapidly consumed and converted to other species. This is shown for the case of n-heptane in Figure 9a. Also presented in this figure is the concentration profile for $C_2H_2$, which is seen to rise to its peak at the flame but then gradually falls.

Figure 9b shows the concentration profiles for $A_1$ and for soot, again for the case of n-heptane combustion. A large amount of aromatic formation occurs at the flame zone, due to the high concentrations of precursor species at this location. The concentration of $A_1$ then increases slightly as it continues to be formed, but eventually decreases as $C_2H_2$ and $C_3H_3$ concentrations drop and PAH growth occurs (for the n-heptane case shown in Figure 9b, the decrease in $A_1$ concentration occurs beyond the time scale shown in the figure). The level of soot rises steadily within the jet plume, as is suggested by the conceptual model of Figure 5.

Addition of the oxygenated fuels (DMM and ethanol) does not alter the shapes of the concentration profiles described above, but does affect the magnitude of the peak concentrations and the rate at which soot is formed.

In the plots of Figure 9, the time scale was chosen to extend to about 0.56 ms. This corresponds to a time of 0.50 ms after reaction takes place in the premixed flame

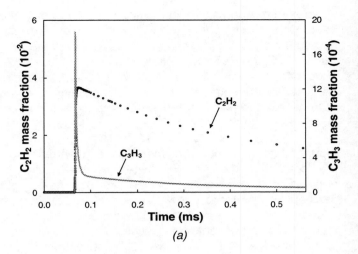

*(a)*

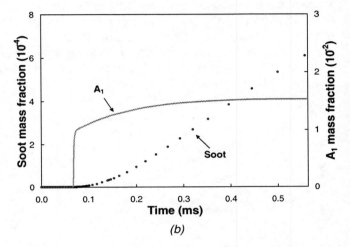

*(b)*

**Figure 9.** Species concentration profiles for the case of n-heptane ($\phi = 4.0$, $T_i = 1000$ K, $P = 10$ MPa). *(a)* Acetylene ($C_2H_2$) and propargyl ($C_3H_3$) and *(b)* benzene ($A_1$) and soot.

(about 0.06 ms, as defined by the point where the $C_3H_3$ concentration reaches its peak). The reasoning behind the use of the 0.5 ms time interval is that it represents the approximate time required for the products of the fuel-rich combustion to travel from the premixed flame to the diffusion flame sheath [29] (about 5 crank angle degrees (CAD) at an engine speed of 1600 rpm). Identified soot concentrations at the location of the diffusion flame can then be used as a predictor for exhaust soot levels, based upon the following assumptions: (1) the diffusion flame oxidizes most but not quite all of the soot created within the plume, and/or (2) quenching of the reaction during the expansion stroke causes the last parcels of injected fuel to produce soot that is not subsequently oxidized.

### Effect of DMM Addition

The gradual replacement of n-heptane by DMM in the numerical model results in notable changes to the peak concentrations of soot precursors at the premixed reaction zone. This subsequently alters the nature of aromatic formation and PAH growth and serves to suppress soot particle inception. Figure 10 shows the peak concentrations of $C_2H_2$, $C_3H_3$ and $A_1$ plotted with respect to oxygen content, as DMM is added to the fuel. As the figure shows, peak $C_2H_2$ and peak $A_1$ concentrations steadily decline with DMM addition. Peak $C_3H_3$ concentrations are not significantly affected at lower oxygen levels, but decrease as oxygen content is increased beyond 10% by mass.

Figure 11 presents the predicted soot concentrations at the location of the diffusion flame (relative to the n-heptane only case), based on the 0.5 ms time-of-flight assumption discussed above. Note that the exact time interval selected would not dramatically affect the relative mass concentration results since soot forms proportionally over time (Figure 12). The data in Figure 11 shows that computed soot concentrations decline in a linear fashion up to a fuel oxygen content of about 17% (corresponding to 40% DMM by volume, or DMM-40). Beyond that point, soot levels continue to decrease with DMM addition until virtually no soot is produced (at a fuel oxygen content between 35% and 40%).

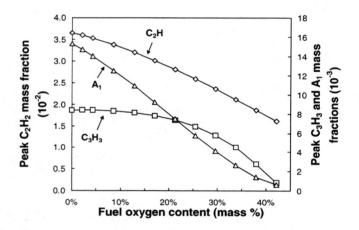

**Figure 10.** Changes to peak soot precursor concentrations due to DMM addition ($\phi = 4.0$, $T_i = 1000$ K, P = 10 MPa).

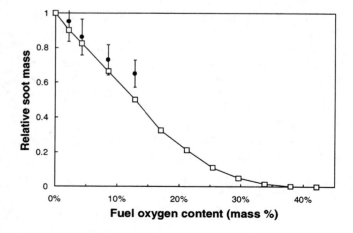

**Figure 11.** Predicted soot mass concentrations at the location of the diffusion flame for DMM addition ($\phi = 4.0$, $T_i = 1000$ K, P = 10 MPa). Solid symbols represent experimental data for PM mass.

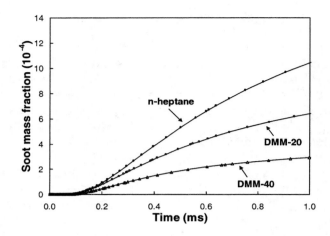

**Figure 12.** Evolution of soot concentrations within the burning jet plume for different levels of DMM addition ($\phi = 4.0$, $T_i = 1000$ K, P = 10 MPa).

Figure 11 also shows the experimental data obtained with the DMM blend fuels using the Cummins B5.9 engine (represented by the solid circles). The numerical simulation reveals a trend similar to that of the experimental data, but would overpredict the amount of PM reduction achieved during actual diesel combustion. This observation is not surprising considering the simplifications made by the model and the fact that additional processes contributing to PM mass (both within the combustion chamber and during exhaust and dilution) are not considered.

<u>Effect of Ethanol Addition</u>

The addition of ethanol to n-heptane in the numerical model produced similar trends to those observed with DMM. Figure 13 shows the results for peak soot precursor levels as oxygen content is increased. The effect of ethanol on peak $C_3H_3$ concentrations is nearly identical to that for DMM. However, based on the model results, ethanol is more effective at reducing peak concentrations of $C_2H_2$ and $A_1$. As a consequence,

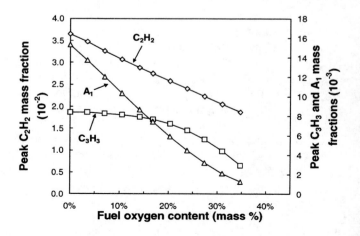

**Figure 13.** Changes to peak soot precursor concentrations due to ethanol addition ($\phi = 4.0$, $T_i = 1000$ K, P = 10 MPa).

ethanol reduced soot concentrations more effectively than DMM for equivalent levels of oxygen addition (Figure 14). Soot production was completely suppressed at a fuel oxygen content around 35%. As was the case with DMM, the magnitude of the soot formation curve decreased with ethanol addition, but the general shape of the curve was not altered. This is illustrated in Figure 15.

Comparison with the experimental results from the B5.9 engine (Figure 16) show that the model again predicts reductions in soot concentrations that are more pronounced than measured PM reductions. However, the model does accurately reproduce the experimental observation that ethanol appears more effective than DMM at lowering PM (as a function of fuel oxygen content). This suggests that the current model is capable of evaluating the relative PM reduction potential of different oxygenated fuels.

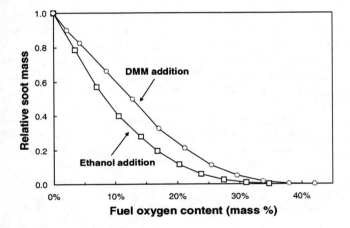

**Figure 14.** Comparison of predicted soot precursor concentrations at the diffusion flame for ethanol addition and DMM addition ($\phi = 4.0$, $T_i = 1000$ K, $P = 10$ MPa).

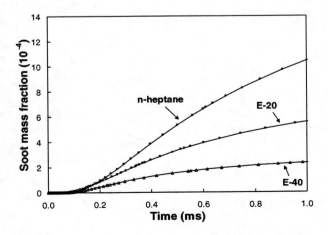

**Figure 15.** Evolution of soot concentrations within the burning jet plume for different levels of ethanol addition ($\phi = 4.0$, $T_i = 1000$ K, $P = 10$ MPa).

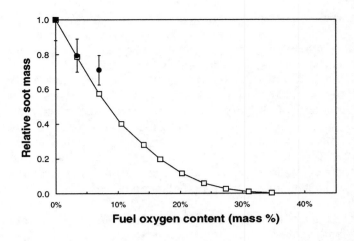

**Figure 16.** Predicted soot mass concentrations at the location of the diffusion flame for ethanol addition ($\phi = 4.0$, $T_i = 1000$ K, $P = 10$ MPa). Solid symbols represent experimental data for PM mass.

### Discussion of the Oxygenate Effect

It is evident from the numerical modeling of oxygenate addition that the key factor affecting soot formation is the nature of the reaction products of the fuel-rich premixed flame. The relative concentrations of species in this zone controls the subsequent processes of aromatic ring formation, PAH growth, and soot particle inception.

A direct effect on product species in the premixed flame zone arises from the difference in pyrolysis products for n-heptane and for the oxygenates DMM and ethanol. N-heptane decomposes primarily via $\beta$-scission and produces small carbon chain species that readily lead to soot precursors. In contrast, DMM lacks any C-C bonds and can only create soot precursors through a more lengthy reaction pathway. Ethanol contains a single C-C bond and cannot immediately form any $C_3$ precursor species, but it is capable of producing $C_2$ species. However, since the C-O bond strength is greater than that of the C-C bond, decomposition of ethanol would in fact tend to yield single carbon species rather than the $C_2$ species that could eventually serve as building blocks for aromatic formation.

An additional, and perhaps more important effect of oxygenate addition is revealed from an investigation of individual species concentrations in the premixed reaction zone. Concentrations of radicals such as O, OH, and HCO are increased, sometimes dramatically, as oxygen addition takes place. The impact that this increase in radical concentrations has on soot formation is two-fold. First, large O and OH concentrations promote oxidation to CO and $CO_2$ within the flame and reduce the amount of carbon available for the production of soot precursor species. Formation of high amounts of HCO results in a similar effect, as HCO readily is converted to CO or $CO_2$ through a single reaction step. Secondly, an increased concentration of radicals, primarily OH, in the post-premixed flame soot formation region, serves to suppress particle inception by oxidizing aromatic species and limiting PAH growth. In fact, one of the main differences between ethanol addition and DMM addition is that peak OH radical concentrations were observed to increase much more dramatically with the addition of ethanol (Figure 17). This is believed to be one of the primary reasons that ethanol is more effective than DMM at reducing soot concentrations. The impact of DMM addition was much greater for the HCO radical, which lowers soot precursor concentrations but would not contribute as significantly in limiting subsequent aromatic ring growth.

Because the production of CO and $CO_2$ limits carbon participation in soot formation reactions, the O/C ratio of a given mixture of oxygenated fuel and air (at a fixed stoichiometry, in this case $\phi = 4.0$) appears to serve as a better parameter than fuel oxygen content for assessing a fuel's soot reduction potential. This illustrated in Figure 18.

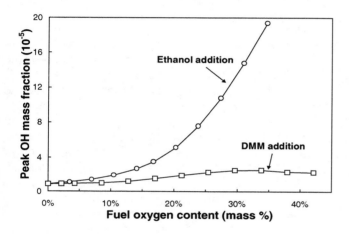

**Figure 17.** Comparison of peak OH concentrations for ethanol addition and DMM addition ($\phi = 4.0$, $T_i = 1000$ K, P = 10 MPa).

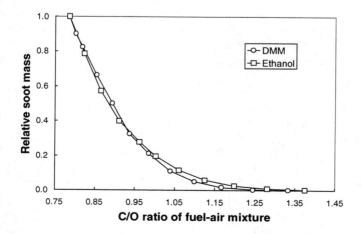

**Figure 18.** Predicted soot precursor concentrations for ethanol and DMM addition, plotted with respect to the C/O ratio of the overall fuel-air mixture ($\phi = 4.0$, $T_i = 1000$ K, P = 10 MPa).

## SUMMARY AND CONCLUSIONS

Experimental results from a Cummins B5.9 diesel operated with oxygenated diesel blends showed that PM reduction levels were influenced largely by the oxygen content of the blend fuel. For the fuels tested, the effect of chemical structure on measured PM mass was observed to be small. Individual modal variations in the effectiveness of oxygenate addition were attributed to the smaller absolute levels of PM at lower load conditions, as well as the contribution of condensed or adsorbed hydrocarbons at those modes. Isotopic tracer tests with ethanol blends revealed that carbon from ethanol did contribute to soot formation, but was about 50% less likely to form soot when compared to carbon from the diesel portion of the fuel.

Numerical modeling results show that oxygenates reduce the production of soot precursors (and therefore soot and PM) through several key mechanisms. The first is due to the natural shift in pyrolysis and decomposition products as oxygen-containing fuels displace the long carbon chains present in conventional diesel fuel. In addition, high radical concentrations produced by the oxygenates in the premixed flame zone promote the oxidation of carbon to CO and $CO_2$, limiting carbon availability for soot precursor formation. An additional effect of high radical concentrations occurs after the premixed flame, where increased OH concentrations limit aromatic ring growth and soot particle inception.

Differences were observed in the two oxygenates evaluated in the numerical modeling (DMM and ethanol). Ethanol showed larger reductions in soot concentrations for equal amounts of oxygen addition. This was believed to be in large part due to the greater amount of OH radicals produced by ethanol addition. Because of the importance of CO and $CO_2$ production in limiting carbon availability for soot formation, the O/C ratio of fuel-air mixtures was found to be a parameter that is well correlated to the ability of an oxygenated fuel to reduce soot particle concentrations.

## ACKNOWLEDGMENTS

Some of this work was performed under the auspices of the U.S. Department of Energy by University of California Lawrence Livermore National Laboratory under Contract No. W-7405-Eng-48. Support was provided by LLNL Laboratory Directed Research and Development grant 01-ERI-007.

## REFERENCES

1. Akasaka, Y., T. Sasaki, S. Kato and S. Onishi. "Evaluation of Oxygenated Fuel by Direct Injection Diesel and Direct Fuel Injection Impingement

Diffusion Combustion Diesel Engines," SAE Technical Paper 901566, 1990.

2. Fleisch, T., C. McCarthy, A. Basu, C. Udovich, P. Charbonneau, W. Slodowske, S.-E. Mikkelsen and J. McCandless. "A New Clean Diesel Technology: Demonstration of ULEV Emissions on a Navistar Diesel Engine Fueled with Dimethyl Ether," SAE Technical Paper 950061, 1995.

3. Wong, G., B. L. Edgar, T. J. Landheim, L. P. Amlie and R. W. Dibble. "Low Soot Emission from a Diesel Engine Fueled with Dimethyl and Diethyl Ether," WSS/CI Paper 95F-162, October 1995.

4. Sorenson, S. C. and S.-E. Mikkelsen. "Performance and Emissions of a 0.273 Liter Direct Injection Diesel Engine Fueled with Neat Dimethyl Ether," SAE Paper 950064.

5. Mikkelsen, S.-E., J. B. Hansen and S. C. Sorenson. "Progress with Dimethyl Ether," International Alternative Fuels Conference, Milwaukee, WI, June 25-28, 1996.

6. Kajitani, S., Z. L. Cheng, M. Konno and K. T. Rhee. "Engine Performance and Exhaust Characteristics of Direct-Injection Diesel Engine Operated with DME," SAE Technical Paper 972973, 1997.

7. Liotta, Jr., F. J. and D. M. Montalvo. "The Effect of Oxygenated Fuels on Emission from a Modern Heavy-Duty Diesel Engine," SAE Technical Paper 932734, 1993.

8. Ullman, T. L., K. B. Spreen and R. L. Mason. "Effects of Cetane Number, Cetane Improver, Aromatics, and Oxygenates on 1994 Heavy-Duty Diesel Engine Emissions," SAE Technical Paper 941020, 1994.

9. Spreen, K. B., T. L. Ullman and R. L. Mason. "Effects of Cetane Number, Aromatics, and Oxygenates on Emissions From a 1994 Heavy-Duty Diesel Engine With Exhaust Catalyst," SAE Technical Paper 950250, 1995.

10. Tsurutani, K., Y. Takei, Y. Fujimoto, J. Matsudaira and M. Kumamoto. "The Effects of Fuel Properties and Oxygenates on Diesel Exhaust Emissions," SAE Technical Paper 952349, 1995.

11. Miyamoto, N., H. Ogawa, T. Arima and K. Miyakawa. "Improvement of Diesel Combustion and Emissions with Addition of Various Oxygenated Agents to Diesel Fuel," SAE Technical Paper 962115, 1996.

12. McCormick, R. L., J. D. Ross and M. S. Graboski. "Effect of Several Oxygenates on Regulated Emissions from Heavy-Duty Diesel Engines," Environmental Science & Technology, vol. 31, no. 4, pp. 1114-1150, 1997.Dec, J. E. "A Conceptual Model of DI Diesel Combustion Based on Laser-Sheet Imaging," SAE Technical Paper 970873, 1997.

13. Miyamoto, N., H. Ogawa, N. M. Nurun, K. Obata and T. Arima. "Smokeless, Low $NO_x$, High Thermal Efficiency, and Low Noise Diesel Combustion with Oxygenated Agents as Main Fuel," SAE Technical Paper 980506, 1998.

14. Maricq, M. M., R. E. Chase, D. H. Podsiadlik, W. O. Siegl and E. W. Kaiser. "The Effect of Dimethoxy Methane Additive on Diesel Vehicle Particulate Emissions," SAE Technical Paper 982572, 1998.

15. Uchida, M. and Y. Akasaka. "A Comparison of Emissions from Clean Diesel Fuels," SAE Technical Paper 1999-01-1121, 1999.

16. Beatrice, C., C. Bertoli, N. Del Giacomo and M.na. Migliaccio. "Potentiality of Oxygenated Synthetic Fuel and Reformulated Fuel on Emissions from a Modern DI Diesel Engine," SAE Technical Paper 1999-01-3595, 1999.

17. Xiao, Z., N. Ladommatos and H. Zhao. "The Effect of Aromatic Hydrocarbons and Oxygenates on Diesel Engine Emissions," Proc. Instn. Mech. Engrs., Vol. 214, Part D, pp. 307-332, 2000.

18. Hess, H. S., A. L. Boehman, P. J. A. Tijm and F. J. Waller. "Experimental Studies of the Impact of Cetaner on Diesel Combustion and Emissions," SAE Technical Paper 2000-01-2886, 2000.

19. Chapman, E., S. V. Bhide, A. L. Boehman, P. J. A. Tijm and F. J. Waller. "Emissions Characteristics of a Navistar 7.3L Turbodiesel Fueled with Blends of Oxygenates and Diesel," SAE Technical Paper 2000-01-2887, 2000.

20. Hallgren, B. E. and J. B. Heywood. "Effects of Oxygenated Fuels on DI Diesel Combustion and Emissions," SAE Technical Paper 2001-01-0648, 2001.

21. Hilden, D. L., J. C. Eckstrom and L. R. Wolf. "The Emissions Performance of Oxygenated Diesel Fuels in a Prototype Diesel Engine," SAE Technical Paper 2001-01-0650, 2001.

22. Yeh, L. I., D. J. Rickeard, J. L. C. Duff, J. R. Bateman, R. H. Schlosberg and R. F. Caers. "Oxygenates: An Evaluation of Their Effects on Diesel Emissions," SAE Technical Paper 2001-01-2019, 2001.

23. Kitagawa, H., T. Murayama, S. Tosaka and Y. Fujiwara. "The Effect of Oxygenated Diesel Fuel Additive on the Reduction of Diesel Exhaust Particulates," SAE Technical Paper 2001-01-2020, 2001.

24. Cheng, A. S. and R. W. Dibble. "Emissions Performance of Oxygenate-in-Diesel Blends and Fischer-Tropsch Diesel in a Compression Ignition Engine," SAE Technical Paper 1999-01-3606, 1999.

25. Cheng, A. S. and R. W. Dibble. "Emissions from a Cummins B5.9 Diesel Engine Fueled with Oxygenate-in-Diesel Blends," SAE Technical Paper 2001-01-2505, 2001.

26. Buchholz, B. A., A. S. Cheng and R. W. Dibble. "Isotopic Tracing of Bio-Derived Carbon from Ethanol-in-Diesel Blends in the Emissions of a Diesel Engine," SAE Techncal Paper (02SFL-12), 2002.

27. Hansen, J. B., B. Voss, F. Joensen and I. D. Siguroardottir. "Large Scale Manufacture of Dimethyl Ether - A New Alternative Diesel Fuel From Natural Gas," SAE Technical Paper 950063, 1995.

28. Hess, H. S., J. Szybist, A. L. Boehman, J. M. Perez, P. J. A. Tijm and F. A. Waller. "Impact of Oxygenated Fuel on Diesel Engine Performance and Emissions," 6th Diesel Engine Emissions Reduction Workshop, San Diego, CA, August 20-24, 2000.

29. Flynn, P. F., R. P. Durrett, G. L. Hunter, A. O. zur Loye, O. C. Akinyemi, J. E. Dec and C. K. Westbrook. "Diesel Combustion: An Integrated View Combining Laser Diagnostics, Chemical Kinetics, and Empirical Validation," SAE Paper 1999-01-0509.

30. Curran, H. J., E. Fisher, P.-A. Glaude, N. M. Marinov, W. J. Pitz and C. K. Westbrook. "Detailed Chemical Kinetic Modeling of Diesel Combustion with Oxygenated Fuels," Fall Meeting of the Western States Section of the Combustion Institute, Irvine, CA, October 25-26, 1999.

31. Curran, H. J., E. M. Fisher, P.-A. Glaude, N. M. Marinov, W. J. Pitz, C. K. Westbrook, D. W. Layton, P. F. Flynn, R. P. Durrett, A. O. zur Loye, O. C. Akinyemi and F. L. Dryer. "Detailed Chemical Kinetic Modeling of Diesel Combustion with Oxygenated Fuels," SAE Technical Paper 2001-01-0653.

32. Kitamura, T., T. Ito, J. Senda and H. Fujimoto. "Detailed Chemical Kinetic Modeling of Diesel Spray Combustion with Oxygenated Fuels," SAE Technical Paper 2001-01-1262.

33. Dec, J. E. "A Conceptual Model of DI Diesel Combustion Based on Laser-Sheet Imaging," SAE Technical Paper 970873, 1997.

34. Golovitchev, V. I., F. Tao and J. Chomiak, "Numerical Investigation of Soot Formation Control at Diesel-Like Conditions by Reduction Fuel Injection Timing," SAE Technical Paper 1999-01-3552.

35. Rente, T., V. I. Golovitchev and I. Denbratt. "Effect of Injection Parameters on Autoignition and Soot Formation in Diesel Sprays," SAE Technical Paper 2001-01-3687.

36. Curran, H. J., P. Gaffuri, W. J. Pitz and C. K. Westbrook. "A comprehensive Modeling Study of n-Heptane Oxidation," *Combustion and Flame* 114:149-177, 1998.

37. Westbrook, C. K., J. Warnatz and W. J. Pitz. *Eighteenth Syposium (International) on Combustion*, The Combustion Institute, pp. 749-767, 1981.

38. Westbrook, C. K., J. Warnatz and W. J. Pitz. *Twenty-Second Syposium (International) on Combustion*, The Combustion Institute, pp. 893-901, 1988.

39. Westbrook, C. K., W. J. Pitz and W. R. Leppard. "The Autoignition Chemistry of Paraffinic Fuels and Pro-Knock and Anti-Knock Additives: A Detailed Chemical Kinetic Study," SAE Technical Paper 912314, 1991.

40. Chevalier, C., W. J. Pitz, J. Warnatz, C. K. Westbrook and H. Melenk. *Twenty-Fourth Syposium (International) on Combustion*, The Combustion Institute, pp. 92-101, 1992.

41. Westbrook, C. K. and W. J. Pitz. Fall Meeting of the Western States Section of the Combustion Institute, Menlo Park, CA, October 18-20, 1993.

42. Wang, H. and M. Frenklach. "A Detailed Kinetic Modeling Study of Aromatics Formation in Laminar Premixed Acetylene and Ethylene Flames," Combustion and Flame 110:173-221, 1997.

43. Wang, H. and M. Frenklach. *Journal of Physical Chemistry* 98:11465-11489, 1994.

44. Wang, H. and M. Frenklach. *Journal of Physical Chemistry* 97:3867-3874, 1993.

45. Edgar, B. L., R. W. Dibble and D. W. Naegeli. "Autoignition of Dimethyl Ether and Dimethoxy Methane Sprays at High Pressures," SAE Technical Paper 971677, 1997.

46. Marinov, N. M, "A Detailed Chemical Kinetic Model for High Temperature Ethanol Oxidation," *International Journal of Chemical Kinetics* 31:183-220, 1999.

47. Shepherd, J. E. "CV" constant volume explosion structure calculation program, http://www.galcit.caltech.edu/EDL/public/codes.html.

48. Kee, Robert J., F. M. Rupley and J. A. Miller. "Chemkin-II: A FORTRAN Chemical Kinetics Packager for the Analysis of Gas-Phase Chemical Kinetics," Sandia Report SAND89-8009B, 1989.

49. Kee, Robert J., F. M. Rupley and J. A. Miller. "The Chemkin Thermodynamic Data Base," Sandia Report SAND87-8215B, 1990.

50. Hess, H. S., J. Szybist, A.L. Boehman, P.J.A. Tijm and F.J. Waller. "The Use of Cetaner for the Reduction of Particulate Matter Emissions in a Turbocharged Direct Injection Medium-Duty Diesel Engine," Seventeenth Annual International Pittsburgh Coal Conference, Pittsburgh, September 11-15, 2000.

## DEFINITIONS, ACRONYMS, ABBREVIATIONS

**0-D**: zero-dimensional
**$^{14}$C**: carbon-14 radioisotope
**AMS**: accelerator mass spectrometry
**CAD**: crank angle degree
**CN**: cetane number
**DEE**: diethyl ether
**DI**: direct-injected
**DME**: dimethl ether
**DMM**: dimethoxy methane
**HC**: hydrocarbon
**HR**: heat release
**NO$_x$**: nitrogen oxides
**NTC**: negative temperature coefficient
**NVOF**: non-volatile organic fraction
**PAH**: polycyclic aromatic hydrocarbon
**PM**: particulate matter
**$\phi$**: fuel-air equivalence ratio

2002-01-1706

# The Possibility of Gas to Liquid (GTL) as a Fuel of Direct Injection Diesel Engine

**Mitsuharu Oguma**
New Energy and Industrial Technology Development Organization, NEDO

**Shinichi Goto**
National Institute of Advanced Industrial Science and Technology, AIST

**Kazuya Oyama**
Ibaraki University

**Kouseki Sugiyama and Makihiko Mori**
Iwatani International Corporation

## ABSTRACT

In this study, engine performances and exhaust emissions characteristics of compression ignition engine fueled with GTL were investigated by comparison with diesel fuel. Diesel engine could be operated fueled with GTL without any special modify for the test engine. With the high cetane number of GTL, the ignition lag was shorter, and the combustion started earlier than that of diesel fuel. Brake thermal efficiency operated with GTL increased at middle load conditions due to incomplete combustion emission such as CO and THC were lower than that of diesel fuel operation. NOx emission with GTL was comparable to diesel fuel, and there was a little decrease at high load. With GTL, soot emission was lower than with diesel fuel at above middle load condition. It seemed to be a reason of soot reduction that there was little sulphur contained in GTL. The brake thermal efficiency with GTL increased with retarding the injection timing, but the decreasing rate regarding the injection timing was lower than with diesel fuel. The NOx and soot emissions with GTL were lower than with diesel fuel at any injection timing.

## INTRODUCTION

Global and urban environmental problems are caused by the rapid increase of carbon dioxide ($CO_2$) and harmful exhaust emissions due to the burning of fossil fuels. Recently, Gas to liquid (GTL) has been attracting attention[1]-[5] like DME[6]-[12] as an alternative fuel.

GTL is a synthetic liquefied fuel produced from natural gas or coal. GTL is a liquid state in the normal temperature and pressure, therefore, conventional infrastructures of fossil fuels can be used. It has high cetane number as well as the other properties are comparable with that of diesel fuel. Therefore, GTL can be an alternative fuel of compression ignition engine. Furthermore, there is little sulfur and aromatic, so clean exhaust emission of diesel engine fueling with GTL can be expected. If the cost is comparable to fossil fuels, GTL might come into the market sooner.

Today, there are 8 GTL refineries in the world. The manufacture capacity of commercial GTL refineries is 200,000 barrels a day, planned manufacture capacity is 600,000 barrels a day since 2010[13]. This is 2.1 % of manufacture capacity of crude petroleum such as 28,000,000 barrels a day. However, manufacture capacity of GTL can be increased because natural gas deposits in the world achieve 90 % of oil deposits. Furthermore, GTL can be saved energy by effective utilization of small- and medium-size gas resources and application to the off-gas emitted from oil fields.

In this study, engine performances and exhaust emissions characteristics of direct injection diesel engine fueled with GTL were investigated by comparison with diesel fuel. And the effects of injection timing on engine performances were also investigated.

## GTL PROPERTIES

Table 1 shows the characteristics of a test GTL fuel using this study compared to diesel fuel and DME. The liquid density, boiling point and flash point of the test GTL are comparable to diesel fuel. Cetane index of GTL is typically high, so its cetane index is higher such as 78 than that of diesel fuel. Although the kinematic viscosity of the GTL is comparable to that of diesel fuel, it seems that the effect of lubrication of GTL on fuel injection system is low, because the sulphur contain in GTL is less than in diesel fuel. Therefore, extremely low sulphur is a good point for reduction of soot emission.

There are no data such as lower heating value and stoichiometric Air-Fuel ratio of test GTL. However, these are calculated by following equations [14], and presented in Table 1.

$$H_{nj} = H_{gj} - 6 \times 4.18605(9h + W) \qquad (1)$$

$$H_{gj} = 4.18605 \times 238.889 \times \left[ (51.916 - 8.792D^2) \right.$$

$$\left. \times \{1 - 0.01(W + A + S)\} + 0.094205 \right] \qquad (2)$$

$$L_0 = \frac{O_{2\,min}}{0.232} = 4.31 \left( 8\frac{c}{3} + 8h + s - o \right) \qquad (3)$$

Where

| | | |
|---|---|---|
| $H_{nj}$ | : Lower heating value | [J/g] |
| $H_{gj}$ | : Higher heating value | [J/g] |
| D | : Density @288K | [g/cm³] |
| S | : Sulphur | [wt.%] |
| W | : Water content | [vol.%] |
| A | : Ash content | [wt.%] |
| h | : Hydrogen content | [wt.%] |

## EXPERIMENTAL APPARATUS AND PROCEDURES

Figure 1 presents a schematic diagram of the engine experiment. The test engine used in this study was a four-stroke, single cylinder, naturally aspirated and direct injection diesel engine (Mitsubishi Motors Corporation DT-30). There was no especial modification for the test engine to operate with GTL. The specifications of the engine and experimental conditions are showed in Table 2. The injection timing was fixed at the point that the brake thermal efficiency was highest operated with diesel fuel. Both, GTL and diesel fuel, operations were conducted at the same injection timing.

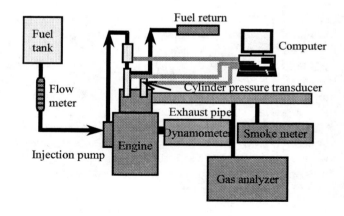

Fig.1 Schematic diagram of the engine experiment

The engine was operated at a constant speed of 1500 rpm. The cylinder pressures, needle lift of injector and the brake specific fuel consumptions were measured for various loads to investigate combustion characteristics of diesel engine with GTL. Also, the exhaust gas analyzer (Best sokki) was used to measure NOx, total hydrocarbon (THC), CO and $CO_2$.

### Table 1 Characteristics of GTL Diesel fuel and DME

| Properties | Unit | GTL | Diesel fuel | DME |
|---|---|---|---|---|
| Chemical structure | | $C_nH_{2.13n}$ | $C_nH_{1.87n}$ | $(CH_3)_2O$ |
| Liquid density | g/cm³ | 0.78 @288 [K] | 0.83 @288 [K] | 0.668 @293 [K] |
| Boiling point | K | 448-633 | 453-643 | 248.1 |
| Vapor pressure | kPa | <0.0001 @313 [K] | | 0.53 @293 [K] |
| Vapor density | air=1 | >5 | | |
| Flash point | K | 369 | 344 | |
| Auto-ignition temperature | K | >493 | 523 | 508 |
| 90% Recovered | K | 605 | 622 | |
| Cetane index | | 78 | 57.8 | >>55 |
| Kinematic viscosity | mm²/s | 3.5 @313 [K] | 3.76 @303 [K] | <1 |
| Lower heating value | J/kg | 46533 | 43200 | 28430 |
| Stoichiometrc A/F | kg/kg | 14.96 | 14.37 | 8.96 |
| Sulphur | % m | <0.0050 | 0.034 | 0 |
| Water content | % v/v | <0.01 | | <0.002 |
| Sediment | % m | <0.01 | 0.01 | 0 |
| Aromatics | % m | <0.1 | | 0 |
| C | % m | 84.9 | 86 | 52.2 |
| H | % m | 15.1 | 14 | 13 |
| O | % m | 0 | 0 | 34.8 |

## Table 2 Engine specifications and experimental conditions

| Bore x Stroke [mm] | 135 x 140 |
|---|---|
| Injection pump plunger [mm dia.] | 12 (Bosch type) |
| Displacement [cm$^3$] | 2003 |
| Original static injection timing [°CA ATDC] | -15.0 |
| Nozzle opening pressure [MPa] | 24.5 |
| Engine speed [rpm] | 1500 |
| Compression ratio | 17.5 |

## RESULTS AND DISCUSSION

## BASIC ENGINE PERFORMANCE

Figure 2 shows the pressure-time history of diesel engine operated with GTL and diesel fuel. For both operations, fuel injection started at about −15 ATDC. The combustion with GTL started earlier at about a few crank angles than that of diesel fuel. It seems that with higher cetane number of GTL, the ignition lag became shorter.

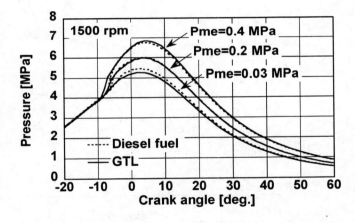

Fig.2 Pressure-time history of diesel engine operated with GTL and diesel fuel

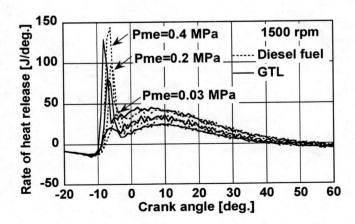

Fig.3 Rates of heat release of diesel engine operated with GTL and diesel fuel

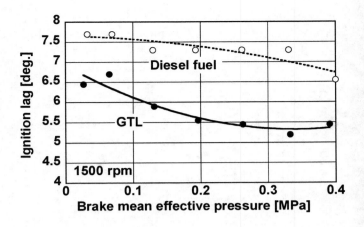

Fig.4 The relationship between ignition lag and brake mean effective pressure (Pme) with GTL and diesel fuel

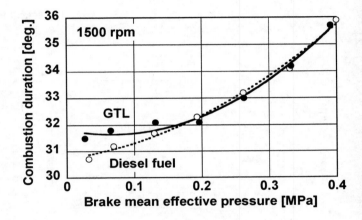

Fig.5 The relationship between combustion duration and Pme with GTL and diesel fuel

Figure 3 shows the rates of heat release calculated from pressure data. The patterns of rates of heat release fueled with GTL were almost same as that with diesel fuel operation. The heat releases of GTL operation were earlier, and the peaks of pre-mixed combustion were lower than that of diesel fuel.

Figure 4 shows the relationship between ignition lag and brake mean effective pressure (Pme) with GTL and diesel fuel. The ignition lags with GTL were about 1 or 2 degrees shorter than with diesel fuel because of higher cetane number of GTL than that of diesel fuel. Therefore, the combustion started earlier and the peak of pre-mixed combustion was lower than that of diesel fuel operation as mentioned in Fig.3.

Figure 5 shows the relationship between combustion duration and brake mean effective pressure with GTL and diesel fuel. The start of combustion was defined as the rise point of the rate of heat release curve, and the combustion end was defined as the crank angle achieved 95 % of total heat release. When the engine was operated fueled with GTL, the combustion duration was longer as compared with diesel fuel at low brake mean effective pressure, and it was comparable to diesel fuel operation at higher brake mean effective pressure.

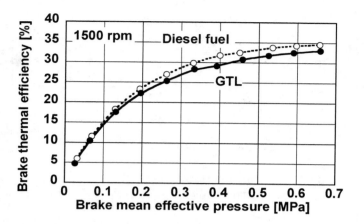

Fig.6 The relationship between brake thermal efficiency and brake mean effective pressure (Pme) with GTL and diesel fuel

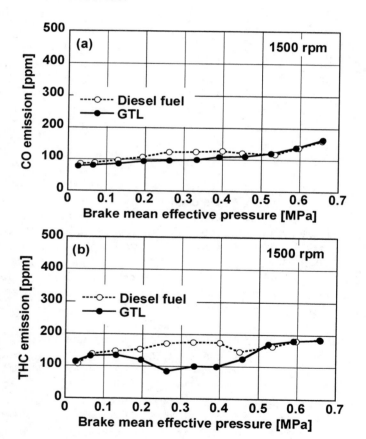

Fig.7 The relationship between (a) CO and (b) THC emissions and brake mean effective pressure with GTL and diesel fuel

Figure 6 shows the relationship between brake thermal efficiency and brake mean effective pressure with GTL and diesel fuel. The brake thermal efficiency of diesel fuel operations achieved 35 % at high load condition. In case of GTL operation, the brake thermal efficiency was

lower a few percentages than of diesel fuel operation at the injection timing of −15 degree ATDC.

Figure 7 shows the relationship between (a) CO and (b) THC emissions and Pme with GTL and diesel fuel. The CO emission fueled with GTL was lower than that of diesel fuel at below Pme=0.5 MPa, and then comparable to diesel fuel operation. The THC emission with GTL was about 60 % of diesel fuel at medium load such as Pme=0.2–0.5 MPa. CO and THC emissions are an indicator of incomplete combustion that is the combustion efficiency with GTL operation was higher for all loads.

Figure 8 shows the relationship between NOx emission and Pme with GTL and diesel fuel. The NOx emission with GTL was comparable to diesel fuel, and there was a little decrease at high load.

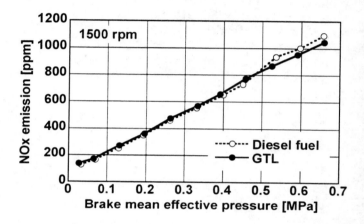

Fig.8 The relationship between NOx emissions and brake mean effective pressure with GTL and diesel fuel

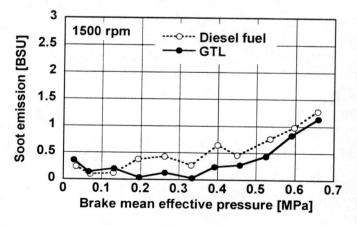

Fig.9 The relationship between soot emissions and brake mean effective pressure with GTL and diesel fuel

Figure 9 shows the relationship between soot emission and brake mean effective pressure with GTL and diesel fuel. The soot emission of GTL was just a little higher than that of diesel fuel at low load, then negligible at medium load and finally emitted about 1.2 BSU (Bosch Smoke Unit). However, the soot emission of GTL was lower than that of diesel fuel at above Pme=0.2 MPa. Soot emission seems to increase by decreasing of ignition lag and increasing of combustion duration. It seems to be one of the reason of growing soot emission at low brake mean effective pressure that the combustion duration fueled with GTL was longer than that of diesel fuel at low brake mean effective pressure. However, it decreased at middle and high load conditions compared to diesel fuel operation in spite of shorter ignition lag and comparable combustion duration with GTL to diesel fuel. That is low sulphur contained in GTL seemed to reduce soot emission.

## EFFECT OF INJECTION TIMING ON ENGINE PERFORMANCE

The ignition lag with GTL is shorter than with diesel fuel because of higher cetane number of GTL. Therefore, the retarding of injection timing seems to be effective for GTL operation. Thus, effect of injection timing on engine performance and exhaust characteristics is discussed in this session.

Figure 10 shows the effect of injection timing on brake thermal efficiency with GTL and diesel fuel. When the engine was operated with diesel fuel, the brake thermal efficiency decreased linearly with retarding the injection timing, however, in case of GTL operation, the decreasing rate of brake thermal efficiency was lower. It seems that the high cetane number of GTL made the ignition lag shorter, and the decreasing rate of brake thermal efficiency with GTL was lower than with diesel fuel.

Figure 11 shows the effect of injection timing on (a)CO and (b)THC emissions with GTL and diesel fuel. CO emission with GTL was a little lower than with diesel fuel for all injection timings. The CO emission increased with retarding the injection timing, however, the increasing rate with GTL was lower than with diesel fuel. The THC emission with GTL was comparable to diesel fuel operation, and the level was from 100 to 130 ppm.

Figure 12 shows the effect of injection timing on (a)NOx and (b)Soot emissions with GTL and diesel fuel. The NOx and soot emissions with GTL were lower than with diesel fuel at any injection timing. The NOx emission with GTL decreased with retarding the injection timing. Particularly, when the injection timing was set at −6 degree ATDC, the NOx emission decreased about 55 % compared to injection timing at −15 degree ATDC maintaining the high brake thermal efficiency such as 32 %, as shown in Fig.10. At this injection timing, the soot emitted only 1.3 BSU.

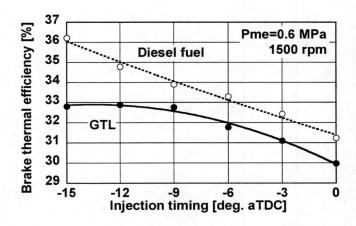

Fig.10 The effect of injection timing on brake thermal efficiency with GTL and diesel fuel

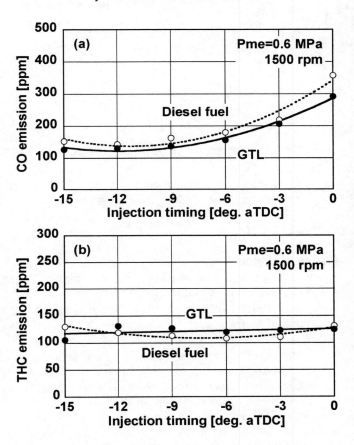

Fig.11 The effect of injection timing on (a) CO and (b) THC emissions with GTL and diesel fuel

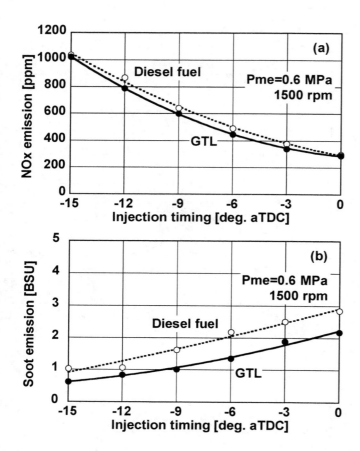

Fig.12 The effect of injection timing on (a) NOx and (b) soot emission with GTL and diesel fuel

## CONCLUSION

Engine performances and exhaust emissions characteristics of compression ignition engine fueled with GTL were investigated by comparison with diesel fuel. The results obtained from this study can be summarized as follows:

1. Diesel engine can be operated fueling with GTL without any special modify for the test engine.
2. With the high cetane number of GTL, the ignition lag is shorter, and the combustion starts earlier than that of diesel fuel.
3. Combustion duration of GTL operation is comparable to diesel fuel operation except at low load condition.
4. Brake thermal efficiency with GTL is lower a few percentages than with diesel fuel operation.
5. NOx emission operated with GTL is comparable to diesel fuel, and there is a little decrease at high load.
6. With GTL, soot emission is lower than with diesel fuel at above middle load condition. It seems to be a reason of soot reduction that there is little sulphur contained in GTL.
7. The brake thermal efficiency with GTL increases with retarding the injection timing, but the decreasing rate regarding the injection timing is lower than with diesel fuel.

8. The NOx and soot emissions with GTL are lower than with diesel fuel at any injection timing. When the injection timing is set at –6 degree ATDC, the NOx emission decreased about 55 % compared to injection timing at –15 degree ATDC maintaining the high brake thermal efficiency and a little increase of soot emission.

## ACRONYMS, ABBREVIATIONS

| | |
|---|---|
| GTL | Gas To Liquid |
| DME | Dimethyl ether |
| THC | Total hydrocarbon |
| CO | Carbon oxide |
| $CO_2$ | Carbon dioxide |
| NOx | Nitrogen oxide |
| ATDC | After piston top dead center |
| Pme (BMEP) | Brake mean effective pressure |
| BSU | Bosch smoke unit |

## REFERENCES

1. Leo L. Stavinoha, Emilio S. Alfaro, Herbert H. Dobbs, Jr. and Luis A. Villahermosa and John B. Heywood, "Alternative Fuels: Gas to Liquids as Potential 21st Century Truck Fuels", SAE Paper, 2000-01-3422, 2000.
2. Kenji IKUSHIMA and Soichi YAMAMOTO, "GAS TO LIQUIDS TECHNOLOGY (in Japanese)", ENGINE TECHNOLOGY, Vol. 1, No. 2, pp. 52-55, 1999.
3. Shiinichi Suzuki, "Gas to Liquids Technology of Natural Gas, High Pressure Gas, Vol38, No.9, 2001.
4. Japan National Oil Corporation (http://www.jnoc.go.jp/c_lng/lngfile7.htm).
5. Shinichi Goto, "Fuels Modified Technology, Development of New Energy Vehicles & Their Materials, pp183-186, 2001.
6. S.C. Sorenson and Svend-Erik Mikkelsen, "Performance and Emissions of a 0.273 Liter Direct Injection Diesel Engine Fuelled with Neat Dimethyl Ether", SAE Paper 950064, 1995.
7. Paul Kapus and Herwig Ofner, "Development of Fuel Injection Equipment and Combustion System for DI Diesels Operated on Dimethyl Ether", SAE Paper 950062, 1995.
8. Theo Fleisch, Chris McCarthy, Arun Basu, Carl Udovich, Pat Charbonneau, Warren Slodowske, Svend-Erik Mikkelsen and Jim McCandless, "A New Clean Diesel Technology : Demonstration of ULEV Emissions on a Navistar Diesel Engine Fueled with Dimethyl Ether", SAE Paper 950061, 1995.
9. Zhili CHEN, Mitsuru KONNO and Shuichi KAJITANI, "Performance and Emissions of DI Compression Ignition Engines Fueled with Dimethyl Ether (1st Report, Performance and Emissions in Retrofitted Engines) (in Japanese)", JSME paper, 64-627-B, pp.383-388, 1998.

10. S. Kajitani, Z. L. Chen, M. Konno and K. T. Rhee, "Engine Performance and Exhaust characteristics of Direct-Injection Diesel Engine Operated with DME, SAE Paper 972973, 1997.

11. Mitsuharu Oguma, Mitsuru Konno, Shuichi Kajitani, Kyung T. Rhee, "A Study of Low Compression-Ratio Dimethyl Ether (DME) Diesel Engine", The ASME Internal Combustion Engine Division, Paper No. 2000-ICE-289, 2000.

12. Mitsuharu OGUMA, Hideyuki MACHIDA, Kiyotaka MINEGISHI, Mitsuru KONNO and Shuichi KAJITANI, "A Study of Low Compression Ratio Dimethyl Ether (DME) Diesel Engine (in Japanese)", JSME Paper, 66-648-B, pp.321-327, 2000.

13. Yukihiro Tsukasaki, "Technical Trend of GTL Fuels for Automobiles(in Japanese)", Journal of Society of Automotive Engineers of Japan, Vol.55, No.5, pp. 67-72, 2001.

14. "Crude petroleum and petroleum products-Determination and estimation of heat of combustion", Japanese Industrial Standards, JIS K 2279, 1993.

2002-01-1707

# Spectroscopic Investigation of the Combustion Process in DME Compression Ignition Engine

**Mitsuharu Oguma and Gisoo Hyun**
New Energy and Industrial Technology Development Organization, NEDO

**Shinichi Goto**
National Institute of Advanced Industrial Science and Technology, AIST

**Mitsuru Konno and Kazuya Oyama**
Ibaraki University

## ABSTRACT

For better understanding of the in-cylinder combustion characteristics of DME, combustion radicals of a direct injection DME-Fueled compression ignition engine were observed using a spectroscopic method. In this initial report, the emission intensity of OH, CH, CHO, $C_2$ and NO radicals was measured using a photomultiplier. These radicals could be measured with wavelength resolution (half-width) as about 3.3 nm. OH and CHO radicals appeared first, and then CH radical emission was detected. After that, the combustion radicals were observed using a high-speed image intensified video camera with band-pass filter. All of radicals were able to observe as images with half-width as 6 or about 10 nm. Rich DME leaked from nozzle was burning at the end of combustion. Therefore, the second light emission of $C_2$ radical after the main combustion was observed. The light of OH emitted all over the DME spray at the combustion started, then the dim light expanded in combustion chamber, so that CH radical emitted at the spray tip. After that the light intensity around the combustion chamber wall became strong with advancing the combustion. The light of NO radical emitted at around the tip of spray, which area seems to be stoichiometric combustion area.

## INTRODUCTION

Global and urban environmental problems are caused by the rapid increase of carbon dioxide ($CO_2$) and harmful exhaust emissions due to the burning of fossil fuels. Available oil deposits are calculated to last only about 42 years, so it is an urgent tusk to convert petroleum fuel to an environmentally safe and renewable new fuel.

Recently, Dimethyl ether (DME) has been attracting much attention as a clean alternative fuel for diesel engine [1-7]. Table 1 shows characteristics of DME compared with propane and diesel fuel. The vapor pressure of DME is comparable to propane, DME liquefy easily with a few pressurization. However, its lower heating value is about 60 % of diesel fuel and viscosity is very low.

The cetane number of DME is high enough to operate a compression-ignition engine, and the thermal efficiency of DME-powered diesel engine is comparable to diesel fuel operation. Low soot free combustion can be achieved without any extra modifications, such as aftertreatment and engine technology. Therefore, DME direct injection (DI) diesel engine is able to use EGR [8] and de-NOx catalyst [9, 10]. However, there are some problems to brake through such as the poor lubricity, NOx emission, low maximum power output and difference of combustion process from diesel fuel.

Since the DME combustion process of DME is different from that of diesel fuel, it is necessary to investigate the combustion process of a DME diesel engine. The combustion mechanism has been elucidating by in-cylinder gas sampling [11, 12], numerical calculation [13, 14] and observation of combustion radicals by using monochromator [15]. However, the light intensity of DME combustion is not strong enough to take some radicals images by using monochromator. In this experiment, DME combustion radicals such as OH, CH, CHO, $C_2$ NO and HCHO were observed from images by using band-pass filters of the halt-width with about 10 nm.

Table 1 Characteristics of DME, propane and diesel fuel

| Property | DME | Propane | Diesel fuel |
|---|---|---|---|
| Chemical formula | $(CH_3)_2O$ | $C_3H_8$ | $C_nH_{1.8n}$ |
| Boiling point [K] | 248.1 | 231 | 453-643 |
| Cetan number | >>55 | - | 57.8 |
| Lower heating value [kJ/kg] | 28430 | 46360 | 43200 |
| Vapor pressure [MPa] at 293[K] | 0.53 | 0.83 | <0.001 |
| Liquid viscosity [cSt] | $0.25(\times10^{-3})$ | - | $2.4-4.6(\times10^{-3})$ |

## EXPERIMENTAL APPARATUS AND PROCEDURES

Figure 1(a) and 1(b) show schematic diagrams of the experimental system for combustion observation using a bottom view type research engine and optical instruments. The quartz window diameter is 47.8 mm, so the in-piston combustion chamber can be observed. The test engine is a four-stroke, single cylinder, naturally aspirated, direct injection diesel engine (Yanmar Diesel Corporation NFD-13K). At normal atmospheric conditions, DME is gaseous state, so pressure-resistant fuel lines were used so that the DME could be pressurized with nitrogen to maintain it in a liquid state. The specifications of the engine and experimental conditions are listed in Table 2.

Due to the lower viscosity of DME, lubricating additive is ordinarily used for reliable operation of a DI DME diesel engine. However, lubricating additives were not used for this work in order to avoid possible optical interference from combustion of these additives.

The spectral emission intensities were measured every crank angles by using photomultiplier (Hamamatsu Photonics 1P28) with a monochromator (MC-30C) as shown in Fig. 1(a), and the spectral emission images were taken by a high-speed video camera (Kodak Ekta Pro HS-4540) with an image intensifier (IMCO PLS-3) at 4500 frames per second using band-pass filters as shown in Fig. 1(b), and the specifications of band-pass filters are shown in Table 3.

When emission intensities of combustion radicals were measured by the photomultiplier, the slit width set at 1mm (the half-width is about 3.3 nm).

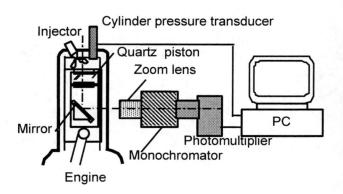

Fig. 1(a) Experimental setup for measuring spectral emission intensities

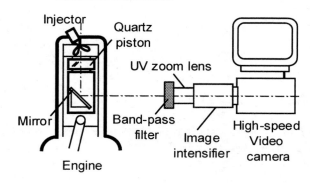

Fig. 1(b) Experimental setup for flame observation

Table 2 Engine specifications and experimental conditions

| Bore × Stroke | 92 × 96 mm |
|---|---|
| Compression ratio | 17.7 |
| Rated output | 8.09 kW (11.0 PS) @2400 rpm |
| Injection pump, plunger | 8 mm dia. (Bosch type) |
| Injector | 0.26 mm hole dia. 4-hole nozzle |
| Original static injection timing | -17.0 °CA ATDC |
| Nozzle opening pressure | 19.6 MPa |
| Pme | 0.4 MPa |
| Engine speed | 960 ±50 rpm |
| Coolant temperature | 363 ±5 K |

Table 3 Specifications of band-pass filters

| Object radicals | Peak waves [nm] | Wave resolution (Half-width) [nm] | Trans-mittancy [%] |
|---|---|---|---|
| OH 306.4 [nm] | 307.1 ±2 | 10 ±2 | 12 |
| CH 431.4 [nm] | 430.0 ±2 | 10 ±2 | 45 |
| C2 516.5 [nm] | 514.0 ±2 | 10 ±1 | 50 |
| CHO 329.8 [nm] | 330.0 ±2 | 12 ±2 | 35 |
| NO 236.3 [nm] | 239.0 ±2 | 10 ±2 | 12 |
| HCHO 395 [nm] | 395.0 ±6 | 6 | - |

## RESULTS AND DISCUSSION

Figure 2(a) shows the pressure history of the DME fueled test engine. This cylinder pressure data is an ensemble average of 45 engine cycles. The rate of heat release was calculated from the cylinder pressure data, and is plotted in Fig. 2(b). If the combustion start is defined as the initial point of heat release, and the combustion end is defined as the crank angle corresponding to 95 % of total heat release, then the combustion duration of this study is 28 deg.

### TIME-DISTRIBUTIONS OF DME COMBUSTION RADICALS

Figure 3 shows emission intensity-time histories of combustion radicals of the DME fueled engine, and Figure 4(a)-(e) show emission intensities and the corresponding rate of heat release. In these figures, the emission intensities are uncalibrated, so they cannot be compared quantitatively. But all combustion radicals could be measured with wavelength resolution half-width of about 3.3 nm.

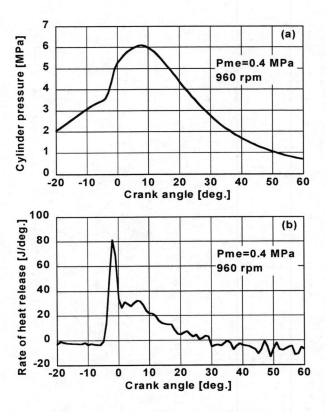

Fig. 2 (a) Pressure-time histories and (b) Rate of heat release of DME diesel engine

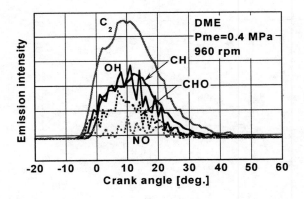

Fig.3 Emission intensity-time history of OH, CH, CHO, $C_2$ and NO radicals of a DI diesel engine fueled with DME

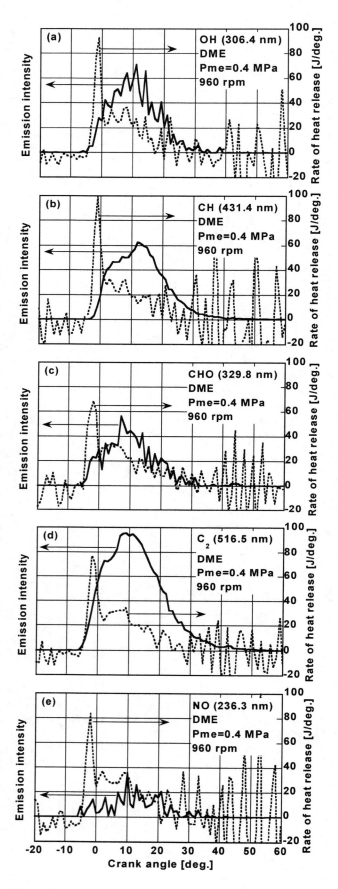

All emission intensities, except NO, increase rapidly at the start of the pre-mixed combustion. NO is detected in very small amounts after top dead center. $C_2$, OH and CHO radicals detect first, and then CH is obtained. At the end of combustion, the emission intensity of OH and CHO drops earlier than that of CH.

Following are some reactions of the DME oxidation process involving with OH radical:

$$CH_3OCH_3 + O = CH_3OCH_2 + OH \cdots\cdots\cdots (2)$$
$$CH_3OCH_2 = CH_3 + CH_2O \cdots\cdots\cdots\cdots (3)$$
$$O + CH_4 <=> OH + CH_3 \cdots\cdots\cdots\cdots (4)$$

Therefore, the OH radical is an important parameter concerning the combustion process of DME.

The CHO radical can be an indicator of incomplete combustion, since it involves a CO reaction pathway for DME oxidation.

Several reactions involved with CH radical are shown as follows:

$$C_2 + OH = CO + CH \cdots\cdots\cdots\cdots\cdots (5)$$
$$CH_2 + O_2 = CH + CO_2 \cdots\cdots\cdots\cdots (6)$$

Thus, the CH radical might be an indicator for active heat release.

From these results, it seems that DME resolve actively to $CH_3$, while CO appears temporary as soon as combustion start, and then active resolution to $CO_2$ occur. According to previous investigations of the formation process of unburned hydrocarbon components from a DME DI diesel engine by in-cylinder gas sampling [11, 12], CO and $CH_3$ ($CH_4$) formed so much immediately after combustion start, and then $CO_2$ increased in response to decrease in those components. These results are in agreement with the spectroscopic results of this study.

$C_2$ radical emission occurred over the entire combustion period. According to the report of in-cylinder gas sampling [11, 12], $C_2$ elements such as $C_2H_4$, $C_2H_6$ and so on, are transient species formed during process of DME combustion. $C_2$ radical emission in hydrocarbon combustion is thought to result from induction of $C_n$ by reaction heat, which is produced through two steps: extraction of hydrogen from hydrocarbon, and polymerization of the unsaturated carbon backbone. Due to its stability, the emission intensity of $C_2$ is very strong.

Fig. 4 Relationship between the (a) OH, (b) CH, (c) CHO, (d) $C_2$, and (e) NO radicals and the rate of heat release of DME DI diesel engine

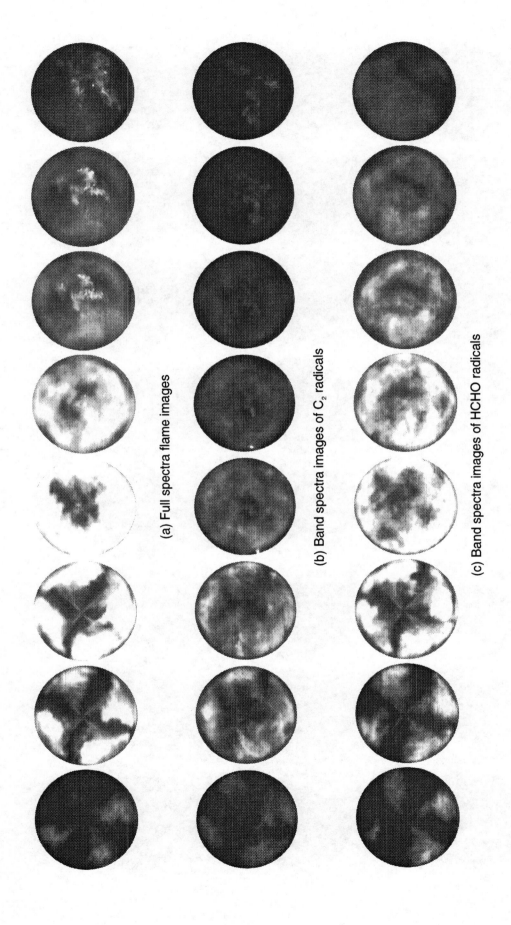

(a) Full spectra flame images

(b) Band spectra images of $C_2$ radicals

(c) Band spectra images of HCHO radicals

Fig.5 Comparison of flame development with (a) the normal images and band spectra images of (b) $C_2$ radical and (c) HCHO radical

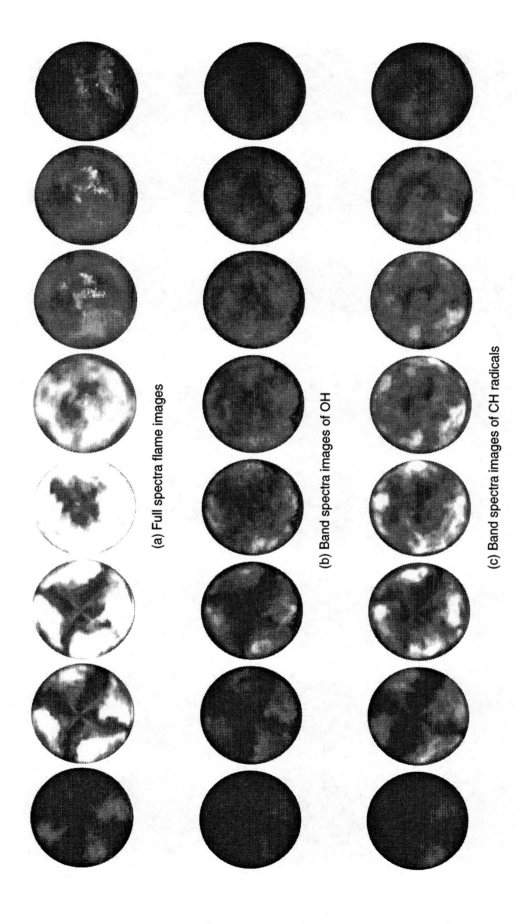

(a) Full spectra flame images

(b) Band spectra images of OH

(c) Band spectra images of CH radicals

Fig.6 Comparison of flame development with (a) the normal images and band spectra images of (b) OH radical and (c) CH radical

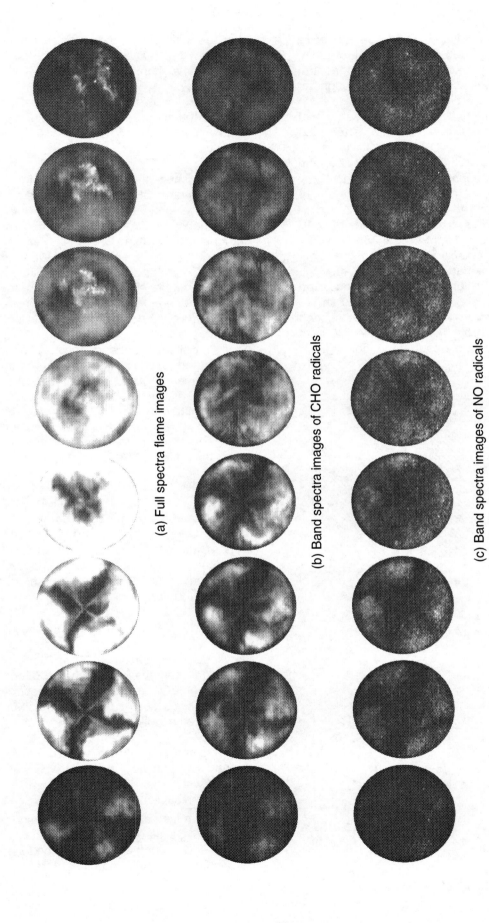

(a) Full spectra flame images

(b) Band spectra images of CHO radicals

(c) Band spectra images of NO radicals

Fig.7 Comparison of flame development with (a) the normal images and band spectra images of (b) CHO radical and (c) NO radical

73

## SPATIAL-DISTRIBUTIONS OF DME COMBUSTION RADICALS

Figure 5, 6 and 7 show the full spectra and band spectra images obtained by using band-pass filters. All of them, they are not always complete band spectra images, but there are some differences in images enough to compare each other. Figure 8 shows the light emission duration calculated from Fig.5-7. The emitting duration of $C_2$ radical was longest of all band spectra. Then next radicals were HCHO, CH, OH, CHO and NO in order. These results correspond to the date of measuring the emission intensities of combustion radicals with DME by using photomultiplier.

Figure 5(a) shows the full spectra flame images of DME combustion process. DME ignites at the tip of the spray near the combustion chamber wall, and then the flame expands to the upstream part of the spray. After the main combustion, second light intensity was occurred near the center of combustion chamber. Figure 9 shows needle lift of injector history of test engine. The second lifting of needle occurred from10 to 20 deg. ATDC with all observation conditions. It seems that the leak of DME from injector was burned.

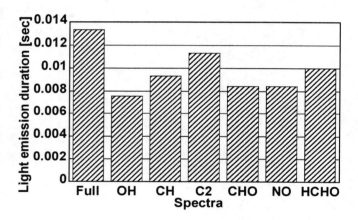

Fig.8 The light emission duration of full flame image and band spectra images

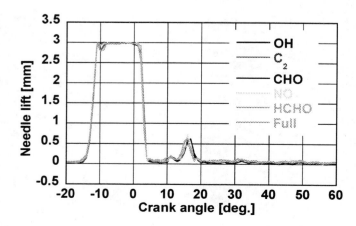

Fig.9 Needle lift of injector time history

Figure 5(b) shows the band spectra image of the $C_2$ radical. The light intensity of $C_2$ radical started to emit at the tip of the DME spray, and the second light emission after the main combustion was also observed. It seems that rich DME leaked from injector was burning at the end of combustion.

Figure 5(c) shows the band spectra image for HCHO radical. It is seen from this figure, the light emitted from all over spray, and the light intensity was strong at the spray tip.

Figure 6 show the band spectra images for (b) OH radicals and (c) CH radicals compared to (a) full spectra image. The light of OH emitted all over the DME spray at the combustion started, then the dim light expanded in combustion chamber. CH radical emitted stronger than OH radical in this experiment. When the combustion occurred, the light of CH radical emitted at the spray tip, then the light observed all over the spray. With advancing the combustion, the light intensity around the combustion chamber wall became strong.

Figure 7 show the band spectra images for (b)CHO radicals and (c) NO radicals compared to (a) full spectra image. In case of CHO, light intensity was more strongly at the lower stream of DME spray. The light of NO radical emitted at around the tip of spray. NO emission was formed actively in the stoichiometric combustion area. Therefore, it seems that the area observed light emission of NO radical is stoichiometric combustion area.

In this study, combustion radicals of DME fueled diesel engine, such as OH, CH, $C_2$, HCHO, CHO and NO were able to observe as images with half-width as 6 or about 10 nm using a high-speed image intensified video camera with band-pass filter. However, these could not be checked as complete band images. It seems that actual spectra of DME combustion need to be scanned in the near future.

## CONCLUSION

In-cylinder combustion characteristics of DME, particularly combustion radicals in direct injection DME diesel engine were observed by spectroscopic method. The results observed with time-distribution could be summarized as follows:

1.  All of combustion radicals such as OH, CH, CHO, $C_2$ and NO can be measured with wave resolution of half-width is about 3.3 nm by using the photomultiplier.

2.  $C_2$ is emitted all over the combustion period. OH and CHO radicals emit first, and then CH is obtained. At the end of combustion, emission intensity of OH and CHO close earlier than that of CH.

3. DME resolve actively to $CH_3$ and CO appears temporarily as soon as the combustion start, and then active resolution to $CO_2$ occur.

The results observed the space-distribution could be summarized as follows:

4. Combustion radicals of DME fueled diesel engine, such as OH, CH, $C_2$, HCHO, CHO and NO were able to observe as images with half-width as 6 or about 10 nm using a high-speed image intensified video camera with band-pass filter. there are some differences in images enough to compare each other.

5. Rich DME leaked from nozzle burns at the end of combustion. Therefore the second light emission of $C_2$ radical after the main combustion is observed.

6. The light of OH emit all over the DME spray at the combustion started, then the dim light expand in combustion chamber, so that CH radical emit at the spray tip. The light intensity around the combustion chamber wall becomes strong with advancing the combustion.

7. The light of NO radical emit at around the tip of spray, which area seems to be stoichiometric combustion area.

## ACKNOWLEDGMENTS

The authors wish to thank Yuichiro Hirayama and Hiroki Matsui, graduate students, Ibaraki University and other students, for their considerable assistance with the experimental work.

## ACRONYMS, ABBREVIATIONS

| | |
|---|---|
| **DME** | Dimethyl ether |
| **DI** | Direct injection |
| **EGR** | Exhaust gas recirculation |
| **CO** | Carbon oxide |
| **CO₂** | Carbon dioxide |
| **NOx** | Nitrogen oxide |
| **ATDC** | After piston top dead center |

## REFERENCES

1. S.C. Sorenson and Svend-Erik Mikkelsen, "Performance and Emissions of a 0.273 Liter Direct Injection Diesel Engine Fuelled with Neat Dimethyl Ether", SAE Paper 950064 (1995)
2. Paul Kapus and Herwig Ofner, "Development of Fuel Injection Equipment and Combustion System for DI Diesels Operated on Dimethyl Ether", SAE Paper 950062 (1995)
3. Theo Fleisch, Chris McCarthy, Arun Basu, Carl Udovich, Pat Charbonneau, Warren Slodowske, Svend-Erik Mikkelsen and Jim McCandless, "A New Clean Diesel Technology : Demonstration of ULEV Emissions on a Navistar Diesel Engine Fueled with Dimethyl Ether", SAE Paper 950061 (1995)
4. Zhili CHEN, Mitsuru KONNO and Shuichi KAJITANI, "Performance and Emissions of DI Compression Ignition Engines Fueled with Dimethyl Ether (1st Report, Performance and Emissions in Retrofitted Engines)", JSME paper, 64-627-B, pp.383-388 (1998), in Japanese
5. S. Kajitani, Z. L. Chen, M. Konno and K. T. Rhee, "Engine Performance and Exhaust characteristics of Direct-Injection Diesel Engine Operated with DME, SAE Paper 972973 (1997)
6. Mitsuharu Oguma, Mitsuru Konno, Shuichi Kajitani, Kyung T. Rhee, "A Study of Low Compression-Ratio Dimethyl Ether (DME) Diesel Engine", The ASME Internal Combustion Engine Division, Paper No. 2000-ICE-289 (2000)
7. Mitsuharu OGUMA, Hideyuki MACHIDA, Kiyotaka MINEGISHI, Mitsuru KONNO and Shuichi KAJITANI, "A Study of Low Compression Ratio Dimethyl Ether (DME) Diesel Engine", JSME Paper, 66-648-B, pp.321-327 (2000), in Japanese
8. Tomoya MORI, Shuichi KAJITANI, Mitsuru KONNO, Mitsuharu OGUMA, Hideto KATO, "The NOx reduction effects by EGR of DME CI engine", JSME Proceedings of Ibaraki District Conference, pp.297-298 (2000), in Japanese
9. Mahabubul Alam, Osamu Fujita, Kenichi Ito, Shuichi Kajitani, Mitsuru Konno, Mitsuharu Oguma, "Comparison of NOx Reduction Performance with NOx Catalyst in Simulated Gas and Actual DME Engine Exhaust Gas", ASME ICE-Vol. 33-1 Book No. G1127A-1999 (1999)
10. M. Alam, O. Fujita, K. Ito, S. Kajitani, M. Oguma, H. Machida, "performance of NOx Catalyst in a DI Diesel Engine operated with Neat Dimethyl Ether", SAE Paper 1999-01-3599 (1999)
11. Mitsuru KONNO, Shuichi KAJITANI and Yoshihiro SUZUKI, "Unburned Emissions from a DI Diesel Engine Operated with Dimethyl Ether", Proceeding of The 15th Internal Combustion Engine Symposium (International), 9935167 (1999)
12. Mitsuru KONNO, Shuichi KAJITANI, Zhili CHEN, Kenji YONEDA, Hiroki MATSUI and Shinichi GOTO, "Investigation of the Combustion Process of a DI CI Engine Fueled with Dimethyl Ether", SAE Paper 2001-01-3504 (2001)
13. Norimasa Iida, "Numerical Calculation Study of Auto-Ignition with Elementary Reactions Fueled with Pre-mixed DME-air gas", JSME Report of RC174 (2000), in Japanese
14. Masaaki OHYA, Kentaro TSUCHIYA, et al, "Oxidation Reaction of Dimethyl Ether", JSME Report of RC174 (2001), in Japanese
15. Mitsuharu Oguma, Gisoo Hyun, Mahabubul Alam, Shinichi Goto, Mitsuru Konno and Kazuya Ooyama, "Combustion Radicals Observation of DME Engine by Spectroscopic Method", SAE Paper 2002-01-0863 (2002)

2002-01-1708

# A Method to Determine Biogas Composition for Combustion Control

**C. Rahmouni**
CReeD

**M. Tazerout and O. Le Corre**
DSEE, Ecole des Mines de Nantes

## ABSTRACT

This paper presents a methodology for a rapid determination of biogas composition using easily detectable physical properties. As biogas is mainly composed of three constituents, it is possible to determine its composition by measuring two physical properties and using specific ternary diagrams. The first part of the work deals with the selection of two physical properties, which are easy and inexpensive to measure, from a group comprising thermal conductivity, viscosity and speed of sound. Then, in the second part, a model to express these properties in terms of ternary composition is presented. It is demonstrated that the composition of a ternary gas mixture can be determined with good precision using the above. The model is applied to specific situations such as the online determination of the lower heating value of biogas without any complicated apparatus like calorimeters or batch techniques (gas chromatographs). The error on the lower heating value and Wobbe index of biogas is less than 1% even when taking into account other constituents not specified in the ternary diagram like oxygen. The effect of small errors in the measurement of physical properties has also been highlighted.

## INTRODUCTION

Biogas is produced by the anaerobic fermentation process (free of oxygen) of animal or vegetal organic matter. This proceeds in three stages (hydrolysis, acid formation and methane fermentation) under the action of certain bacteria. This process of waste decomposition or biodegradation is natural and inevitable. Production of biogas depends on many factors: kind of feed stock and its water content, temperature, pH, etc [1]. Depending on the feed stock used for its production, biogas can also contain variable quantities of water, hydrogen sulfide and oxygen. Biogas generally contains more than three constituents. However, biogas composition can be represented by its most important constituents (see Table 1) [1]. Biogas has been used in stationary engines as well as in mobile engines.

| Gas constituent | Variation range (%) | Average content (%) |
|---|---|---|
| Methane | 30-65 | 45 |
| Carbon dioxide | 20-40 | 35 |
| Nitrogen | 5-40 | 15 |
| Hydrogen | 1-3 | 1 |
| Oxygen | 0-5 | 1 |
| Hydrogen sulfide | 0-0.01 | 0.003 |

**Table 1: Variations of biogas constituents**

When biogas is used to run an internal combustion engine, changes in composition can have serious effects on performance. For instance, there can be power output fluctuations due to variation in the heating value ($LHV$ can vary at the extreme from 10 to 25 $MJ/m^3$). Thus, a measurement of biogas composition becomes essential in order to control engine operation for optimal performance. The Wobbe index, which expression is defined in next section, is another important characteristic of gaseous fuels (it can vary from 10 to 30 $MJ/m^3$) [2]. It is an important criterion dealing with the interchangeability of gaseous fuels in engines. Changes in fuel gas composition do not induce noticeable variations in air-to-fuel ratio and combustion velocity when the Wobbe index remains almost constant. Two gases of different compositions but same Wobbe index can be substituted for each other to provide globally the same volumetric energy content. Thus, knowing the Wobbe index of biogas is important in order to upgrade biogas close to natural gas quality [3]. From a technical point of view, the most important difference between biogas and natural gas is that the Wobbe index for natural gas is twice the value of biogas. Only gases with a similar Wobbe index can substitute each other and allow distribution of biogas in the natural gas network.

Different techniques exist for finding the composition of biogas. Chromatography, which is a batch technique, is one of them. However, it requires skilled technical operation and maintenance (it sets a high accuracy against a high cost, a slow response time and a regular calibration). Other measurement methods for determining the calorific value [4,5] and the Wobbe index [5,6] are generally limited to natural gases.

Ternary mixtures, which are representatives in most cases of biogas, can be represented in a ternary diagram [8]. The aim of this work is to establish a simple procedure for rapid evaluation of the composition of biogas and thereby estimate its Wobbe index and *LHV*. The procedure uses two physical properties of biogas and the $CH_4$-$CO_2$-$N_2$ ternary diagram. Then in this paper, the following points are tackled:

- Selection of adequate and suitable physical properties to determine ternary compositions,
- Description of two mathematical models based on two best couples of physical properties,
- Application of the models to real gases.

## MODELS FOR THE CALCULATION OF PHYSICAL AND COMBUSTION PROPERTIES

This section gives available correlations for calculation of viscosity and thermal conductivity [9]. The theoretical formula for the calculation of speed of sound [10], as well as formulae for the lower heating value, the specific gravity and the Wobbe index of a gas [11], are also given. These formulae are used for the display of constant physical property curves in ternary diagrams and the elaboration of the mathematical models to determine ternary compositions and evaluate its uncertainties. All the simulations have been carried out through the software, Matlab.

PHYSICAL PROPERTIES

Viscosity

The viscosity of mixtures can be determined by the following relation:

$$\eta_m = \sum_{i=1}^{n} \frac{y_i \eta_i}{\sum_{j=1}^{n} y_j \Phi_{ij}} \tag{1}$$

Where,

$$\Phi_{ij} = \frac{[1+(\eta_i/\eta_j)^{1/2}(M_j/M_i)^{1/4}]^2}{[8(1+M_i/M_j)]^{1/2}} \tag{2}$$

$\phi_{ji}$ is determined by exchanging indices.

$\eta_i$ is the viscosity of the pure constituent *i*. $y_i$ and $y_j$ are the volume composition of constituent *i* and *j*.

Conductivity

In a form analogous to the theoretical relation for mixture viscosity, an empirical relation can be used for the mixture conductivity:

$$\lambda_m = \sum_{i=1}^{n} \frac{y_i \lambda_i}{\sum_{j=1}^{n} y_j A_{ij}} \tag{3}$$

Where $\lambda_m$ is the thermal conductivity of the mixture and $\lambda_i$ the thermal conductivity of the pure constituent *i* (can be calculated, when data is not available, thanks to viscosity). $A_{ij}$ could be expressed as:

$$A_{ij} = \frac{\varepsilon[1+(\lambda_{tr_i}/\lambda_{tr_j})^{1/2}(M_i/M_j)^{1/4}]^2}{[8(1+M_i/M_j)]^{1/2}} \tag{4}$$

Where *M* is the molecular weight (g/mol) and $\lambda_{tr}$ is the thermal conductivity due to translational energy of molecules (monatomic gases). $\varepsilon$ is a numerical constant near unity. Thus, the relation for estimating mixture viscosity is also applicable to thermal conductivity by simply substituting $\lambda$ for $\eta$. In this approximation, to determine $\lambda_m$, one needs data giving the pure component thermal conductivity and viscosity or obtains the ratio of translational thermal conductivity from Eq. (5):

$$\frac{\lambda_{tr_i}}{\lambda_{tr_j}} = \frac{\Gamma_j[\exp(0.0464T_{r_i}) - \exp(-0.2412T_{r_i})]}{\Gamma_i[\exp(0.0464T_{r_j}) - \exp(-0.2412T_{r_j})]} \tag{5}$$

Where,

$$\Gamma = 210 \left( \frac{T_c M^3}{P_c^4} \right)^{1/6} \tag{6}$$

Where $\Gamma$ is the reduced, inverse thermal conductivity, $[W/(m.K)]^{-1}$, $T_c$ is the critical temperature (K), *M* is the molecular weight (g/mol), and $P_c$ is the critical pressure (in bars).

Speed of sound

The speed of sound *c* for a mixture of gases is given by

$$c^2 = \frac{\gamma_m RT}{M_m} \tag{7}$$

Where $\gamma_m$ is the ratio of specific heats for the gas mixture, $R = 8314.5$ J/mole.K is the universal gas constant, $T$ is the temperature, and $M_m$ is the molecular weight of the mixture. The specific heat ratio and molecular weight of the mixture are based on the mole fraction weighted quantities such that

$$c^2 = \left( \frac{\sum_i y_i C_{pi}}{\sum_i y_i C_{vi}} \right) \frac{RT}{\sum_i y_i M_i} \qquad (8)$$

$C_{pi}$ (respectively $C_{vi}$) is the specific heat at constant pressure (respectively at constant volume) for component $i$. The specific heats are dependent upon temperature [8].

Specific gravity

$$d_m = \sum_i y_i d_i \qquad (9)$$

$d_i$ is the specific gravity of component $i$.

COMBUSTION PROPERTIES

Combustion properties (here, the lower heating value and the Wobbe index) are calculated using the composition of biogas estimated by ternary diagrams.

Lower heating value

The following relation gives the lower heating value of a gas mixture

$$LHV_m = \sum_i y_i LHV_i \qquad (10)$$

$LHV_i$ is the lower heating value of component $i$.

Wobbe index

The following relation gives the Wobbe index of a gas mixture

$$W = \frac{LHV_m}{\sqrt{d_m}} \qquad (11)$$

$d_m$ is the specific gravity of the gas mixture.

SELECTION OF ADEQUATE PHYSICAL PROPERTIES

It is clear from Table 1 that the major constituents of biogas are $CH_4$, $CO_2$ and $N_2$. Hence, a $CH_4$-$CO_2$-$N_2$ ternary diagram can be used to represent physical properties. Examples of such diagrams are given in Figure 1&2. Simulations have been carried out to determine how binary groups of physical properties vary with composition of biogas. Only ternary gases have been considered and many ternary compositions can be found for one value of the physical property used.

Constant physical property curves are then plotted in the diagram together with polynomial fit of the data. When these polynomials are straight lines, from the point of intersection of two lines (corresponding to two physical properties), the ternary gas composition can be determined. In order to have a good intersection, these two physical properties have to be quite distinct from each other. That is to say that two curves having almost the same trend do not give a very good point of intersection. This is the reason why the two physical properties, which will be chosen to elaborate the models, should be well identified.

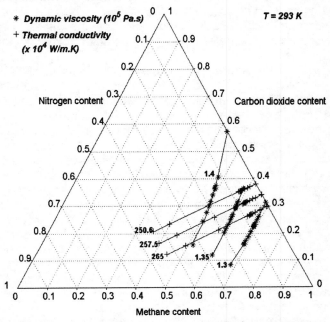

Figure 1: Constant thermal conductivity ($\lambda$) and viscosity ($\eta$) curves

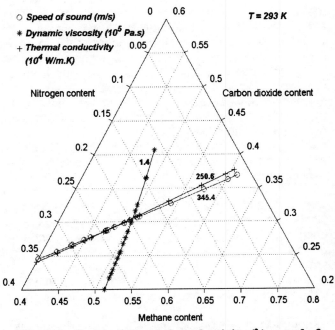

Figure 2: Constant thermal conductivity ($\lambda$), speed of sound ($c$) and viscosity ($\eta$) curves within the biogas domain

Besides, considerations such as accuracy, easy of measurement and cost have to be taken into account. At last, it is also important to look into the errors on the measurement of these properties.

The physical properties displayed in Figure 1&2 are the thermal conductivity ($\lambda$) together with the speed of sound ($c$) and the dynamic viscosity ($\eta$) for $CH_4$-$CO_2$-$N_2$ ternary gases. Figure 2, representative of biogas domain, is a zoom in of Figure 1. These diagrams show that, by constructing for instance two graphs of constant speed of sound or constant thermal conductivity together with constant viscosity, in the $CH_4$-$CO_2$-$N_2$ (biogas) ternary diagram, the exact composition of the ternary gas mixture can be determined from the point of intersection.

Prior to the expression of these properties in terms of ternary compositions, it is necessary to consider carefully the different combinations of two physical properties in order to get the best precision. First, among the three major constituents in biogas (methane, carbon dioxide and nitrogen), only two are independent. Second, some of the physical properties considered are sensitive to one or two constituents of biogas. Thus, viscosity is sensitive to carbon dioxide and nitrogen content variations and less sensitive to methane content variations (constant viscosity curve close to constant methane line). The speeds of sound of methane, carbon dioxide and nitrogen have not close values, which means that speed of sound of biogas is sensitive to all constituents. Carbon dioxide and methane have specific absorption band in the infrared region. All these considerations allow us to combine judiciously two physical properties. In order to demonstrate the above, ternary diagrams are used.

In Figure 2, the relatively small sensitivity of viscosity to methane is well shown (constant viscosity curve close to the methane constant line at 50%) whereas speed of sound does not show such dependence. It can be also noticed that thermal conductivity together with speed of sound of biogas cannot be used without a loss of precision. These two physical properties seem to be correlated between each other. At last, using a $CO_2$ or methane infrared sensor would allow getting (in combination with speed of sound or thermal conductivity) the biogas composition with a good precision. Using the measurement of dynamic viscosity instead of an infrared sensor would give the same result.

The choice for the physical properties depends on whether the physical properties are well correlated with each other or not. In other words, if two constant curves are distinct, it means that the two physical properties are not correlated with each other. This results in a better accuracy. The choice of physical properties also depends on the accuracy of measurement and its cost.

Although thermal conductivity with speed of sound is not an adequate combination, it can be noticed in Figure 3 that thermal conductivity has a different behavior (in the ternary diagram) according to the temperature at which it is measured. This phenomenon makes thermal conductivity, a unique physical property to determine the ternary composition. Measuring thermal conductivity at high temperatures results in better accuracy in determining methane content. Indeed, the thermal conductivity curve at 300°C is closer to the constant methane line (like for viscosity in Figure 2) in the ternary diagram than the one at 0°C. Thus, error in the measurement of thermal conductivity will have less influence on the methane content.

In order to select the adequate easy-to-measure properties according to their precision and cost, an error of ± 1% on the measurement of speed of sound and thermal conductivity and an error of ± 2% in volume for a $CO_2$ infrared sensor are applied. The effect of such errors can be viewed on the ternary diagrams in Figure 4 for a 50% $CH_4$, 30% $CO_2$ and 20% $N_2$ biogas.

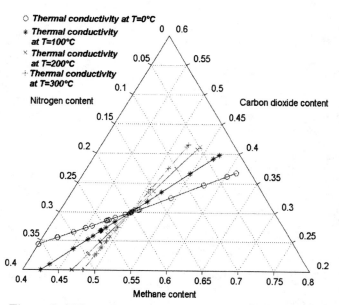

**Figure 3: Effect of temperature on thermal conductivity**

Using speed of sound does not seem to be the best choice. The ± 1% error on its measurement results in a bigger variation in methane content than when using thermal conductivity for instance (area of error delimited by a parallelogram in the ternary diagram). Using an infrared $CO_2$ sensor (or an infrared methane sensor) seems to be the best choice but such sensors are less accurate for high concentration (sensors based on low-resolution spectral absorption). The spanning in concentration can be larger than the ± 2% error in volume. Besides, these sensors are more expensive than thermal conductivity sensors, which are of common use. As a conclusion, a compromise between accuracy, cost and easy-to-measure properties results in the use of thermal conductivity sensors operating at two different temperatures or thermal conductivity together with an infrared $CO_2$ sensor.

A synthesis of the ternary diagram comparing the easiness of measurement, the price and the precision is given in Table 2.

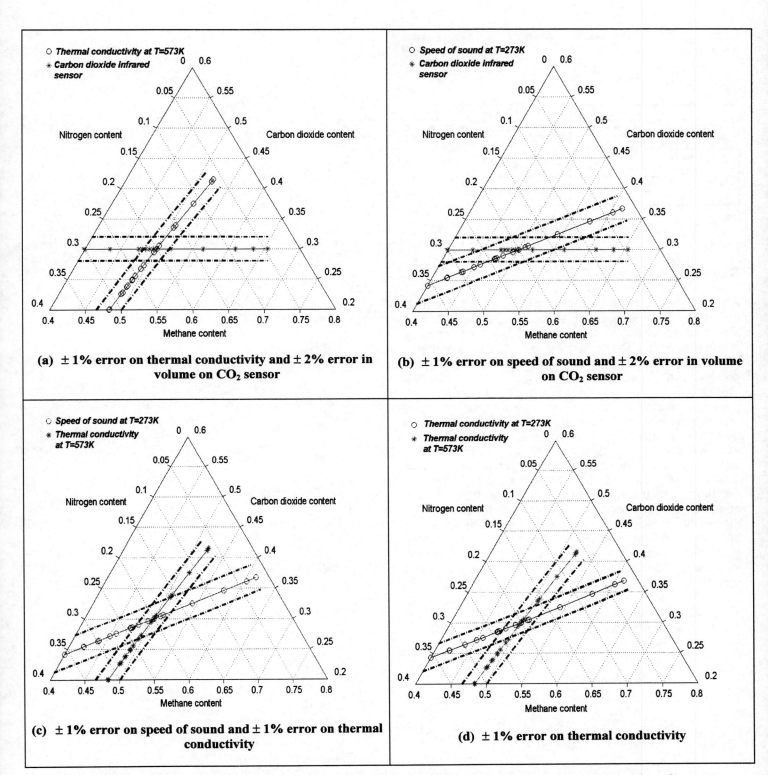

**Figure 4: Effect of errors on thermal conductivity, speed of sound and CO₂ content for the determination of ternary compositions**

Taking speed of sound together with conductivity will result in a cheap sensor against bad precision whereas a combination of speed of sound or thermal conductivity together with viscosity sets a high accuracy against a high cost and the difficulty to measure it. Finally, speed of sound or conductivity together with an infrared CO₂ sensor (or an infrared methane sensor) seems to be an accurate cost-effective solution.

Using an infrared CO₂ sensor instead of a methane infrared sensor may be a better choice. At high concentrations, both CO₂ and methane infrared sensors are not very accurate. The error can exceed ± 2% in volume (up to ± 10%). In that case, the estimation of methane content with the ternary diagram is less influenced by such variations in carbon dioxide content. If we look at figure 4a, an error of ± 5% instead of ± 2%

in the Carbon dioxide content measurement would imply an error of around ± 3% on the methane content determination, which is better than the ± 5% uncertainty when using an infrared sensor.

**Table 2: Comparison of all physical properties**

|  | Precision | Cost | easiness |
|---|---|---|---|
| Viscosity | good | expensive | difficult |
| Conductivity | good | cheap | easy |
| Speed of sound | Bad | cheap | easy |
| CO₂ or CH₄ sensor | good | cheap | easy |

## RESULTS: MATHEMATICAL MODELS

A method to express each gas content as a function of two physical properties is described in [7]. The relations are found to be linear. Thus, by measuring two physical properties of a specific ternary gas, the gas contents ($CH_4$, $CO_2$ and $N_2$) can be determined with a good precision. In Figure 1 and Figure 2, physical properties are calculated for a temperature of 293 K. In fact, physical properties are temperature-dependant and pressure-dependant but pressure has little influence on thermal conductivity over a large range. However, the models should at least take into account the dependence of physical properties with temperature. This would be interesting when using thermal conductivity as unique physical property to determine the composition. Thus, whatever the ambient temperature is, the ternary composition can be calculated by means of thermal conductivity measured at ambiance and at any other temperature above ambiance. Here, this has no importance because thermal conductivity is used at two specific temperatures (273 K and 573 K).

The relations below describe the intersection of two straight lines in an equilateral triangle (Figure 5) [7,8].

$$X_1 = X_{10} + \left(1 - \frac{a_1(\phi_1,T)}{\sqrt{3}}\right)\left(\frac{b_1(\phi_1,T) - b_2(\phi_2,T)}{a_2(\phi_2,T) - a_1(\phi_1,T)}\right) - \frac{b_1(\phi_1,T)}{\sqrt{3}} \quad \text{(12)}$$

$$X_2 = X_{20} + \frac{2a_1(\phi_1,T)}{\sqrt{3}}\left(\frac{b_1(\phi_1,T) - b_2(\phi_2,T)}{a_2(\phi_2,T) - a_1(\phi_1,T)}\right) + \frac{2b_1(\phi_1,T)}{\sqrt{3}} \quad \text{(13)}$$

$$X_3 = 1 - X_1 - X_2 \quad \text{(14)}$$

$\phi_1$ and $\phi_2$ designate the two physical properties used to determine any triplet of gases thanks to the general equations. Here, $X_1 \equiv CH_4$ (methane), $X_2 \equiv CO_2$ (carbon dioxide) and $X_3 \equiv N_2$ (nitrogen) and the sum of $X_i$ must be equal to 100%. $X_{10}$ corresponds to the lower bound of the $X_1$ axis (here 0.4) and $X_{20}$ corresponds to the lower bound of the $X_2$ axis (here 0.2).

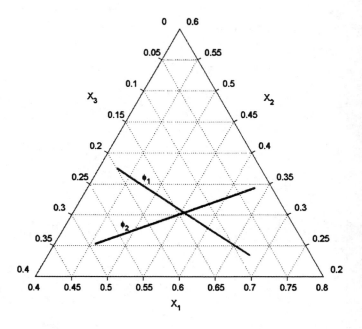

**Figure 5: Domain of application of the models**

For a $CH_4$-$CO_2$-$N_2$ diagram, the coefficients for thermal conductivity at 573 K and an infrared $CO_2$ sensor are given below:

$$a_1(\lambda_{573}) = \alpha_1 \cdot \lambda_{573}{}^2 + \beta_1 \cdot \lambda_{573} + \gamma_1 \quad \text{(15)}$$

$$b_1(\lambda) = \alpha_1' \cdot \lambda_{573} + \beta_1' \quad \text{(16)}$$

$$a_2(\%CO_2) = 0 \quad \text{(17)}$$

$$b_2(\%CO_2) = \frac{\sqrt{3}}{2} CO_2 \quad \text{(18)}$$

Where,

$\alpha_1 = -410.35275$
$\beta_1 = 40.04133$
$\gamma_1 = 0.3254$
$\alpha_1' = -36.5384$
$\beta_1' = 1.97102$

These relations are valid for:

500 W/m.K < $\lambda_{573}.10^4$ < 650 W/m.K
20% < $CO_2$ < 40%

If thermal conductivity is the unique measured property, Eq. (14) and Eq. (15) are kept whereas Eq. (16) and Eq. (17) are modified into Eq. (18) and Eq. (19) at 273 K.

$$a_2(\lambda_{273}) = \alpha_2 \cdot \lambda_{273}{}^2 + \beta_2 \cdot \lambda_{273} + \gamma_2 \quad \text{(18)}$$

$$b_2(\lambda_{273}) = \alpha_2' \cdot \lambda_{273} + \beta_2' \quad \text{(19)}$$

Where,

At 273K:

$\alpha_2 = 4380.1668$

$\beta_2 = -198.3734$

$\gamma_2 = 2.6386$

$\alpha'_2 = -66.5223$

$\beta'_2 = 1.5645$

These relations are valid for:

500 W/m.K < $\lambda_{573}.10^4$ < 650 W/m.K, T = 573 K
210 W/m.K < $\lambda_{273}.10^4$ < 270 W/m.K, T = 273 K

These relations are valid over the whole ternary diagram for biogas only (according to composition in Table 1) and every combination of speed of sound, thermal conductivity and infrared $CO_2$ determination could be used.

The ignition temperature of methane is about 773-813 K, which is higher than the temperature at which thermal conductivity is measured. As a consequence, biogas will not ignite at that temperature. Besides, thermal conductivity sensors that operate up to 600 K can be found in the market.

At last, it can be seen in Figure 3 that the greater the difference of temperature at which thermal conductivity is measured, the lower the error.

### Errors on the models

These models have been built numerically by correlating the coefficients with the physical properties. In order to evaluate the numerical errors of the models, three different ternary gases are considered in Table 3,

|  | GAS 1 | GAS 2 | GAS 3 |
|---|---|---|---|
| CH$_4$ | 65 % | 50 % | 49.5 % |
| CO$_2$ | 35 % | 30 % | 19.5 % |
| N$_2$ | 0 % | 20 % | 31 % |

**Table 3: Composition of the three gases to be tested**

The input data for the models are given in Table 4,

|  | GAS 1 | GAS 2 | GAS 3 |
|---|---|---|---|
| $\lambda_{573K}$ | 613.5 | 567.8 | 546.8 |
| $\lambda_{273K}$ | 235.1 | 230.9 | 218.2 |
| CO$_2$ | 35 % | 30 % | 19.5 % |

**Table 4: Thermal conductivity ($10^4$W/mK) and $CO_2$ content of gases from Table 3**

These input data give us, the ternary composition that is compared to the "real" ternary gases in Table 3. The results are given in Table 5 and Table 6 below,

|  | GAS 1 | GAS 2 | GAS 3 |
|---|---|---|---|
| CH$_4$ | 64.86 % | 49.92 % | 49.2 % |
| CO$_2$ | 35 % | 30 % | 19.5 % |
| N$_2$ | 0.14 % | 20.08 % | 31.3 % |

**Table 5: Ternary compositions obtained by the $\lambda$ (573 K) / $CO_2$ model**

|  | GAS 1 | GAS 2 | GAS 3 |
|---|---|---|---|
| CH$_4$ | 64.81 % | 49.93 % | 49.15 % |
| CO$_2$ | 34.77 % | 30.08 % | 19.23 % |
| N$_2$ | 0.42 % | 19.99 % | 31.62 % |

**Table 6: Ternary compositions obtained by the $\lambda$ (573 K) / $\lambda$ (273 K)**

The error in both models is below 0.7% on the determination of methane content. Both models describe very well the whole domain where biogas is applicable.

Finally, from the composition, important characteristics of biogas such as the lower heating value and the Wobbe index can be estimated.

### Sensitivity of the models to temperature variation

Temperature can be very important when using physical properties sensitive to it (speed of sound, thermal conductivity). In our case, thermal conductivity is measured at a higher temperature than the ambiance. Temperature has to be stabilized. If we consider that temperature can be stabilized within ± 5 K, the error can be seen in the ternary diagram displayed in Figure 5. Measuring thermal conductivity at 573 K (stars in the diagram) and at 568 or 578 K (dashed lines in the diagram) seem to have negligible effect on the physical property and on the model: the iso-thermal conductivity lines are coinciding. We obtain the same results for thermal conductivity at 273 ± 5 K. The errors resulting from uncertainties on temperature stabilization are negligible.

### APPLICATION TO REAL GASES

Lower heating value and Wobbe index calculated using the models of this work are compared extensively for biogas with amounts of oxygen and hydrogen. The results are summarized in Table 9 and Figure 7. In Table 7 are summarized some typical biogas compositions that could be supplied in SI engines. The first gas takes into account all the components usually found in biogas like hydrogen and oxygen. The other three gases are meant to bring up the uncertainties of the model due to oxygen and hydrogen in the gas. The last two ones represent typical binary and ternary gases.

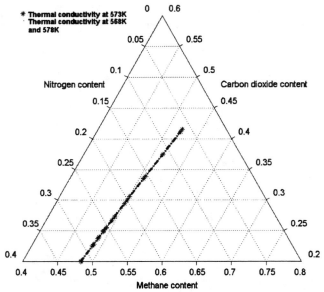

**Figure 6: effect of temperature on thermal conductivity**

|  | GAS 1 | GAS 2 | GAS 3 | GAS 4 | GAS 5 | GAS 6 |
|---|---|---|---|---|---|---|
| **CH$_4$** | 47 % | 60 % | 59 % | 57 % | 60 % | 50 % |
| **CO$_2$** | 35 % | 35 % | 30 % | 30 % | 40 % | 30 % |
| **N$_2$** | 15 % | 0 % | 10 % | 10 % | 0 % | 20 % |
| **H$_2$** | 1 % | 0 % | 1 % | 3 % | 0 % | 0 % |
| **O$_2$** | 2 % | 5 % | 0 % | 0 % | 0 % | 0 % |

**Table 7: Biogas composition**

The gases represented in Table 7 have all more than three components. Ternary diagram should not apply to that kind of gases. However, the extra components have low content compared to the major ones (methane, carbon dioxide and nitrogen). Therefore, in order to calculate the errors generated by extra amounts of hydrogen or nitrogen, the thermal conductivity at 573 K and 273 K will be calculated according to the compositions given in Table 7. These values of thermal conductivity are real values (the ones that could be measured by thermal conductivity sensors). These values will be inputs data for the mathematical models (Table 8). Thus, there will be errors on the determination of methane content, carbon dioxide content and nitrogen. Consequently, there will be errors on the calculation of LHV and the Wobbe index.

|  | GAS 1 | GAS 2 | GAS 3 | GAS 4 | GAS 5 | GAS 6 |
|---|---|---|---|---|---|---|
| **$\lambda_{573K}$** | 567.5 | 599.2 | 608.7 | 625.4 | 593.5 | 567.8 |
| **$\lambda_{273K}$** | 229.1 | 232.4 | 241.4 | 250.2 | 227 | 230.9 |
| **CO$_2$** | 35 % | 35 % | 30 % | 30 % | 40 % | 30 % |

**Table 8: Thermal conductivity ($10^4$W/mK) and CO$_2$ content of gases from Table 2**

## Errors due to sensor measurements

As seen in Figure 4, applying an uncertainty on sensor measurements ($\pm$ 1% for thermal conductivity and $\pm$ 2% for CO$_2$ determination by an infrared sensor) delimits an area (parallelogram) where the maximum of error given by the two models considered are:

- $\pm$ 4% error on the calculation of the lower heating value and the Wobbe index of biogas (with an infrared sensor),
- $\pm$ 4.5% error when using thermal conductivity sensor.

These calculations were performed for GAS 6 of Table 7. It is easy to find thermal conductivity sensors with higher precision:

- For $\pm$ 0.5% of uncertainty on thermal conductivity, the relative error on the lower heating value can vary from −2.5% to 2.2% for both methods,
- For $\pm$ 0.25% of uncertainty on thermal conductivity, the relative error on the lower heating value can vary from −1.71% to 1.4% with the infrared CO$_2$ sensor and from −1.3% to 1% when using two thermal conductivity sensors.

In one hand, a higher precision on the measurement of the carbon dioxide content of biogas by the mean of an infrared sensor will improve the uncertainties of the method using such sensors but in the other hand, its cost will greatly increase. It is difficult to have good precision with such sensors because they are less sensitive to high concentrations of the gas that is measured. Improving the precision of such sensors can considerably decrease the uncertainties on the determination of biogas composition.

Finally, it would be cheaper and more accurate to use two thermal conductivity sensors in order to determine biogas composition, since thermal conductivity can be measured with good accuracy by simple sensors available in the market.

## Errors due to components such as H$_2$ and O$_2$

In order to evaluate the error due to components such as hydrogen and oxygen, six biogas compositions have been considered among which four of them contain oxygen and hydrogen. One gas contains both hydrogen and oxygen and the other three gases contain either hydrogen, either oxygen. The results are summarized in Table 9 and Figure 7.

The lower heating value and the Wobbe index of biogas calculated are given with less than 1% relative error when no hydrogen is present in the gas. When only oxygen is present in the gas (at a maximum content of 5%), the relative error on the LHV is less than 0.5% (for both couple of sensors). The average oxygen content in most biogas is about 1% only. In that case, the error decreases at less than 0.2%.

| | CONDUCTIVITY AT T=573K /CO₂ SENSOR | | | | | | CONDUCTIVITY AT T=573K/CONDUCTIVITY AT T=273K | | | | | |
|---|---|---|---|---|---|---|---|---|---|---|---|---|
| | LHV | | | W | | | LHV | | | W | | |
| | TERNARY DIAGRAM | REAL | %D | TERNARY DIAGRAM | REAL | %D | TERNARY DIAGRAM | REAL | %D | TERNARY DIAGRAM | REAL | %D |
| GAS 1 | 18.20 | 16.98 | 7.2 | 16.82 | 17.29 | -2.7 | 18 | 16.98 | 6.1 | 16.87 | 17.29 | -2.4 |
| GAS 2 | 21.65 | 21.54 | 0.55 | 22.63 | 22.41 | 1 | 21.64 | 21.54 | 0.47 | 21.63 | 22.41 | 0.99 |
| GAS 3 | 22.37 | 21.28 | 5.1 | 23.86 | 22.64 | 5.4 | 22.16 | 21.28 | 4.1 | 23.83 | 22.64 | 5.3 |
| GAS 4 | 24.31 | 20.78 | 17 | 26.27 | 22.22 | 18.2 | 23.66 | 20.78 | 13.9 | 26.21 | 22.22 | 17.9 |
| GAS 5 | 21.36 | 21.54 | -0.8 | 21.95 | 22.16 | -0.9 | 21.32 | 21.53 | -0.98 | 21.94 | 22.16 | -0.97 |
| GAS 6 | 17.92 | 17.95 | -0.17 | 18.58 | 18.61 | -0.2 | 17.92 | 17.95 | -0.1 | 18.58 | 18.61 | -0.2 |

**Table 9: % Deviation on the LHV (MJ/m³) and Wobbe index (MJ/m³) of biogas containing $H_2$ and $O_2$ according to the model used**

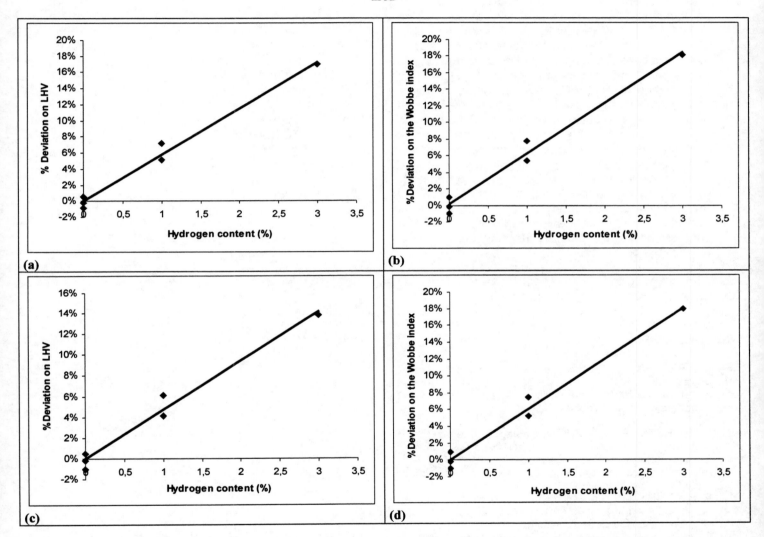

**Figure 7: %Deviation on LHV and the Wobbe index. (a) and (b) are deduced from the couple λ (T=573 K) / CO₂ infrared sensor; (c) and (d) are deduced from the couple λ (T=573 K) / λ (T=273 K)**

This weak error is mainly due to the fact that oxygen and nitrogen have close thermal conductivity while their lower heating value is equal to zero. When only hydrogen is present, the error increases proportionally to the hydrogen content in the gas (Figure 7). In that case, the model is not valid anymore and it is needed to add a correction (for instance, by adding an another easy-to-measure physical property) that could take into account the effect of $H_2$ in the gas. The proportional error caused by hydrogen is due to its very high thermal conductivity compared to that of methane, carbon dioxide and nitrogen. Some of the physical properties that are sensitive to hydrogen are speed of sound and thermal conductivity.

## CONCLUSION

1. A procedure has been developed to estimate the composition of biogas, using two physical properties. Two sets of properties have been chosen. It has been demonstrated that thermal conductivity at two temperatures and a $CO_2$ infrared sensor are the best ones for estimating biogas composition with minimum error.
2. Two models, using the chosen properties, have been outlined. The first model uses the combination of thermal conductivity at 573 K together with $CO_2$ infrared determination. The second model uses the combination of thermal conductivity at 573 K and 273 K. The models can be applied to determine variations in the lower heating value and the Wobbe index of biogas resulting in a better control of engine operation for optimal performance.
3. The method is valid for gases containing methane, carbon dioxide and nitrogen. Gases with oxygen can also be taken into account. In that case, the error on the lower heating value and the Wobbe index of biogas is less than 1%, even with noticeable amount of oxygen (up to 5%).
4. At last, a cheap sensor using thermal conductivity can be easily used to predict variations of *LHV* and the Wobbe index of biogas.
5. The models can be generalized to estimate the composition using measured thermal conductivity at any two temperatures.

Experiments have to be carried out to check the validity of the model and the viability of the sensor.

## REFERENCES

1. M. BEREZA « Economical and Technical Biogas Valorization in SI Engines » MS Degree, 1999 (in French)
2. J. KLIMSTRA « Interchangeability of Gaseous Fuels – The Importance of the Wobbe-Index », SAE paper 861578, 1987
3. K. ERIKSEN et al. « The Upgrading of Biogas to be Distributed Through The Natural Gas Network - Environmental Benefits, Technology and Economy» The Academician Journal of the Chartered Academy, Denmark 1999
4. R. R. THURSTON et al. « Measuring Volume and Calorific Value to Determine an Energy Value of Supplied Gas », UK Patent Application GB 2340945, 1999
5. O. FLORISSON et al. « Rapid Determination of Wobbe Index Meter for Combustion Control », J. Phys. E: Sci. Instrum., 1989
6. H. VERBEEK et al. « Development of a New Wobbe Index of Natural Gas », Meas. Sci. Technol., 1993
7. C. RAHMOUNI et al. « Method for the determination of energetic properties of gaseous fuels », Patent n°0110197, 2001
8. J. A. CORNELL « Experiments with mixtures », Second edition, 1990, John WILEY & Sons, Inc.
9. R. C. REID et al. « The properties of GASES & LIQUIDS », Fourth edition, 1987.
10. D. H. STAELIN et al. « Electromagnetic Waves », Prentice Hall International Editions, 1994, Prentice-Hall International, Inc.
11. « BT104 – Gaseous Combustibles and Combustion Principles », Gaz de France, 1993 Edition

## DEFINITIONS, ACRONYMS, ABBREVIATIONS

| | | |
|---|---|---|
| $P$ | Pressure | [Pa] |
| $T$ | Temperature | [K] |
| $R$ | Universal gas constant | [J/(mol.K)] |
| $C_p$ | Specific heat at constant Pressure | [J/(mol.K)] |
| $C_v$ | Specific heat at constant Volume | [J/(mol.K)] |
| $\eta$ | Dynamic viscosity | [Pa.s] |
| $\lambda$ | Thermal conductivity | [W/(m.K)] |
| $c$ | Speed of sound | [m/s] |
| $M$ | Molecular weight | [g/mol] |
| $LHV$ | Lower Heating Value | [MJ/m$^3$] |
| $W$ | Wobbe Index | [MJ/m$^3$] |
| $d$ | Specific gravity | [-] |
| $\alpha, \alpha'$ | Constants | [-] |
| $\beta, \beta'$ | Constants | [-] |
| $\gamma$ | Constant | [-] |

Subscript:

| | | |
|---|---|---|
| m | mixture | [-] |
| i | indices for component i | [-] |
| c | critical | [-] |
| r | reduced | [-] |
| tr | translational | [-] |

2002-01-1710

# Hydrogen/Oxygen Additives Influence on Premixed Iso-Octane/Air Flame

**A. Sobiesiak, C. Uykur and D. S-K. Ting**
Mechanical, Automotive and Materials Engineering, University of Windsor

**P. F. Henshaw**
Civil and Environmental Engineering, University of Windsor

## ABSTRACT

The effects of the addition of small amounts of molecular and atomic hydrogen/oxygen on laminar burning velocity, pollutant concentrations, and adiabatic flame temperatures of premixed, laminar, freely propagating iso-octane flames are investigated using CHEMKIN kinetic simulation package and a chemical kinetic mechanism at different equivalence ratios. It is shown that hydrogen/oxygen additives increase the laminar burning velocities. Increased hydroxyl (OH) concentrations resulted in reduced carbon monoxide (CO) emissions in every stoichiometric ratios investigated. Additives also increased the adiabatic flame temperature of iso-octane/air combustion, thereby causing increased $NO_x$ concentrations for all additives at all stoichiometries.

## INTRODUCTION

In recent years, an increase has been observed in the number of studies on numerical modeling of the chemical kinetics of combustion. There is continued interest in developing a better understanding of the oxidation of larger hydrocarbon fuels over a wide range of operating conditions. This interest is motivated by the need to improve the efficiency and performance of currently operating combustors, and to reduce the production of pollutant emissions generated in the combustion process. Recent modeling studies of premixed systems have helped in the development of detailed chemical kinetic mechanisms describing iso-octane oxidation [1]. The ability to simulate different fuel/oxidant configurations directly in different operating conditions is the main advantage of using chemical kinetics. The difficulty posed by the use of detailed kinetic mechanisms is related to the computational demands of the large set of nonlinear equations used to simulate the reactions [2]. With existing computational capacities, either fuel chemistry or combustor geometry must be simplified to perform analyses with detailed chemical kinetic mechanisms. However, studies have shown that both the simplified and detailed kinetic

mechanisms can yield essentially identical simulation results, at least for the combustion properties of concern, if the associated reactions are included [3].

Iso-octane (octane number = 100) is a primary reference fuel (PRF) used with n-heptane (octane number = 0) to define the octane reference scale for fully blended gasoline. Recently, Ranzi et al. [4] and Glaude et al. [5] have generated reduced chemical kinetic mechanisms to describe PRF oxidation at low, intermediate, and high temperatures. Roberts et al. [2] have used a semi-detailed chemical mechanism, along with a knock submodel, to simulate iso-octane ignition in a Cooperative Fuels Research (CFR) engine. Different methods have been proposed to produce reduced mechanisms for different problem types and fuels [6]. This process has been automated via proper grouping of the reactions involved [7].

The addition of hydrogen to larger hydrocarbon fuels has been found to result in improved ignitibility, flame holding, emissions, and reactivity [8-12]. The problems associated with handling and storing of hydrogen are still unresolved challenges. For the mean time, point-of-use electrolysis of water is believed to be a practical alternative. Molecular oxygen, a by-product of water electrolysis, when used in conjunction with hydrogen, has been shown to reduce the harmful emissions from burning methane/air mixtures, in addition to further burning velocity enhancement [13]. The production of radicals and oxidizing species such as atomic oxygen (O) and ozone ($O_3$), which are known to enhance the overall combustion process and increase the rate of reaction of the hydrocarbon fuel, can be achieved by corona discharges. Experimental results have demonstrated that improvements in the fuel consumption and emission characteristics, of both gasoline and diesel engines, are possible in the presence of positive corona discharges localized in the pre-combustion air stream [14]. Atomic hydrogen is known as a highly reactive substance and space propulsion systems utilizing atomic hydrogen as propellant have been proposed [15]. Although combustion of iso-octane and hydrogen has been performed in different studies separately, the physicochemical mechanisms of the combustion of the mixture of these fuels have not been reported in any detail. While it might be assumed that the combustion characteristics of a mixed fuel lie between those of the two fuel components, the complex and nonlinear nature of chemical kinetics and multi-component diffusion make it difficult to predict the performance of the mixed fuels [16].

## MODELING TECHNIQUE AND KINETIC MECHANISM

Sandia's steady-state, laminar, one-dimensional flame code PREMIX [17], which uses a hybrid time-integration/Newton iteration technique to solve the steady-state comprehensive mass, species, and energy conservation equations in freely propagating flame configurations, is used in conjunction with TWOPNT, a boundary value problem solver and a transport property preprocessor, in all the calculations performed in this study. CHEMKIN-III [17], a gas-phase interpreter, is used to perform the chemical reaction mechanism calculations.

Peters' [18] iso-octane mechanism comprises of 200 reactions and 56 chemical species and can be used to describe low and high temperature auto-ignition, fuel decomposition and oxidation, as well as the formation of soot precursors. The mechanism is described extensively in reference [18]. However, nitrogen chemistry is not included in the mechanism. In order to simulate NOx emissions in this study, nitrogen containing species and reactions with their rate coefficients have been taken from the GRI mechanism [19]. The number of imported reactions from GRI is 108 and the number of species is 19. The final combined mechanism consists of 308 reactions and 75 species. It is found that the inclusion of nitrogen containing species and the corresponding reactions has a negligible effect on burning velocities, temperature profiles, and CO/HC emissions.

Transport properties of the species in Peters' [18] iso-octane mechanism are obtained by using EGLib - Multicomponent Transport Software [20]. For the nitrogen containing species exported from GRI, their transport property database [19] is used. All the mechanism parameters used in this study such as reaction rates, species involved, and their chemical and transport properties are applied without any modification.

The problem environment is defined by setting the initial and boundary conditions. The initial flow rate estimation of the fuel/oxidizer mixture is set equal to 0.04 [g/cm²-sec] according to published measurements of stoichiometric iso-octane/air burning velocities [21-24]. The pressure is fixed at 1.0 atm and the unburnt mixture temperature is set at 298 K. The additional boundary condition required for the flow solution is supplied by fixing the point at which the temperature is 400 K. The distance to this point is calculated from the initial

temperature profile estimated by the software. Since it is known from previous studies [24] that the total reaction zone of the premixed, laminar, freely propagating, stoichiometric iso-octane/air flame is about 0.4 cm, the calculation domain is started 2 cm before the flame region, and the total length of the calculation domain is chosen as 12 cm. The initial temperature profile estimation, which is required to start the iteration, is made according to the recent study of VanMaaren et al. [25]. Similar estimations are made for the product and peak intermediate mole fractions. Results of the first simulation step are then used as the temperature profile and species maximum intermediate concentration estimation for the next step.

## VALIDATION OF THE SOLUTION ALGORITHM

The solution algorithm is first tested for accuracy by comparing the burning velocity results at different equivalence ratios of pure iso-octane/air mixtures with published values. Since the solution algorithm (adiabatic, steady, one-dimensional, planar, premixed flame) used in this study is independent of external effects such as those of aerodynamics, heat and radical exchanges, the calculated burning velocity $S_u^0$ of the flame represents only its reactivity in the presence of diffusive transport. Thus an agreement between the calculated and measured burning velocity is a good indication on the accuracy of the kinetic mechanism used [26].

Extensive data on the laminar burning velocity have built-up over the years for different fuel/oxidizer combinations. Values of laminar burning velocity, $S_u^0$, reported in the literature have been characterized by substantial scatter. This scatter is partly due to the fact that no experiments can directly generate an ideal one-dimensional, planar, adiabatic, steady, unstrained, laminar flame for which $S_u^0$ is well defined. Earlier techniques, such as the Bunsen burner and constant volume bomb, are believed to have accuracy limitations associated with non-adiabaticity and flame stretch effects coupled with non-uniform diffusion. Burning velocity measurements of those hydrocarbon fuels which are liquids at room temperature are difficult, since the fuel must be gasified and homogeneously mixed with an oxidizer before combustion [27]. It is thus not surprising to see that there is a lack of experimental data for iso-octane.

Comparison results with different measurement techniques [21-23] are shown in Figure 1. Bunsen burner [23] and constant volume bomb [22] results have been directly obtained from measurements without any adjustments. Davis and Law [21], on the other hand, used a nonlinear extrapolation algorithm to predict the unstretched laminar burning velocities from their counterflow twin flame measurements. The higher burning velocities obtained by other techniques may be partly due to flame stretching but, the efficacy of nonlinear extrapolation technique to eliminate these effects is still questionable.

The burning velocity is defined as the speed, relative to and normal to the flame front, with which unburned gas moves into the front and is transformed to products under laminar flow conditions [28]. In this study it is simply the velocity of gas flow at the adiabatic boundary of the flame (flow velocity at $x$ = 0.0 distance). The mechanism used in this study has been optimized particularly for stoichiometric iso-octane/air mixtures and has been shown also to be in good agreement with experimental data for lean mixtures [18]. It should be also mentioned that the equivalence ratio range of reliable iso-octane mechanisms can only be extended at the expense of computational time, since those species produced in very lean and very rich mixtures and their associated reactions have to be included along with their chemical, transport and rate parameters. It is possible to find iso-octane mechanisms composed of 990 different chemical species and 4060 reactions [29].

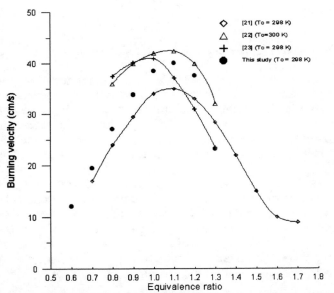

**Fig. 1. Calculated and measured laminar burning velocities of iso-octane/air mixtures as a function of equivalence ratio ($T_0$ = 298 K, $P_0$ = 1 atm).**

Figure 1 compares measured burning velocities of iso-octane/air mixtures at standard conditions as a function of equivalence ratio. The calculated burning velocities are generally in good agreement with well-accepted experimental results from the literature. The simulated burning velocities agree particularly well with Davis and Law's data for lean and rich mixtures, and with Gülder's data for stoichiometric and slightly rich mixtures. However, the present mechanism predicts slightly lower burning velocities than those measured by Gülder [22] and by Gibbs and Calcote, [23] for lean mixtures. The equivalence ratio at the point of maximum burning velocity is predicted to be approximately $\phi \cong 1.1$, where $\phi$ is the fuel/air equivalence ratio. Most experimental results show that under standard atmospheric conditions

the maximum burning velocity of iso-octane/air mixture occurs slightly rich ($\phi \approx 1.1$). The exception is the measured values of Gibbs and Calcote, where the burning velocity peaked at $\phi = 0.98$, this is not in accord with the general behavior of similar hydrocarbon fuels [22]

Laminar burning velocity predictions for higher initial pressures and higher initial temperatures are compared with Bradley's approximations, which are made empirically with an expression of the form:

$$S_u = S_{u,d} \left( \frac{T_0}{T_d} \right)^{\alpha} \left( \frac{P_0}{P_d} \right)^{\beta} \qquad (1)$$

where $S_{u,d}$ is the unstretched laminar burning velocity at the datum temperature ($T_d$) and pressure ($P_d$), $\alpha$ and $\beta$ are constants which depend on $\phi$ [24]. The kinetic simulation results obtained in this study and Bradley's empirical results are shown in Figure 2 for stoichiometric iso-octane/air mixtures at an initial temperature of $T_0$ = 358 K over a range of initial pressures ($P_0$) from 1 to 10 atm. The effect of pressure on burning velocity is simulated in the mechanism by providing Troe type pressure-dependent reaction parameters for some of the reactions. Results showed this approximation is sufficient for simulating pressure dependency of the burning velocity.

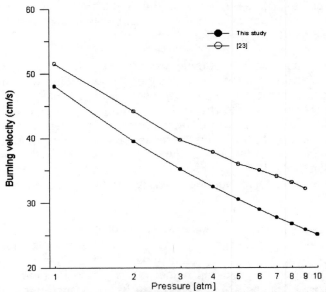

**Fig. 2. Comparison of laminar burning velocity predictions via chemical kinetic simulation with empirical values [24] at elevated pressures ($T_0$ = 358 K, $\phi$ = 1.0).**

## RESULTS AND DISCUSSION

Numerical solutions have been obtained for the following mixtures: 1) pure iso-octane/air, 2) 5% molecular hydrogen + 95% iso-octane + air, 3) 10% molecular hydrogen + 90% iso-octane + air, 4) 5% molecular hydrogen + 95% iso-octane + corresponding amount of molecular oxygen + air, 5) 10% molecular hydrogen +

90% iso-octane + corresponding amount of molecular oxygen + air, 6) 5% atomic hydrogen + 95% iso-octane + air, 7) 10% atomic hydrogen + 90% iso-octane + air, 8) 5% atomic hydrogen + 95% iso-octane + corresponding amount of atomic oxygen + air, and 9) 10% atomic hydrogen + 90% iso-octane + corresponding amount of atomic oxygen + air. The initial concentrations of these mixtures are summarized in Figure 3.

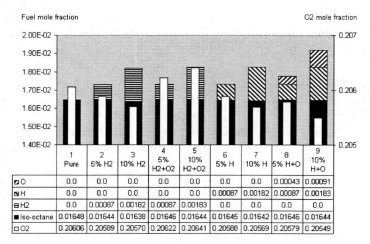

Fuel mole fraction                                    O2 mole fraction

| | 1 Pure | 2 5% H2 | 3 10% H2 | 4 5% H2+O2 | 5 10% H2+O2 | 6 5% H | 7 10% H | 8 5% H+O | 9 10% H+O |
|---|---|---|---|---|---|---|---|---|---|
| O | 0.0 | 0.0 | 0.0 | 0.0 | 0.0 | 0.0 | 0.0 | 0.00043 | 0.00091 |
| H | 0.0 | 0.0 | 0.0 | 0.0 | 0.0 | 0.00087 | 0.00182 | 0.00087 | 0.00183 |
| H2 | 0.0 | 0.00087 | 0.00182 | 0.00087 | 0.00183 | 0.0 | 0.0 | 0.0 | 0.0 |
| Iso-octane | 0.01648 | 0.01644 | 0.01638 | 0.01646 | 0.01644 | 0.01645 | 0.01642 | 0.01646 | 0.01644 |
| O2 | 0.20606 | 0.20589 | 0.20570 | 0.20622 | 0.20641 | 0.20588 | 0.20569 | 0.20579 | 0.20549 |

**Fig. 3. The reactant composition: mole fractions of iso-octane/H₂/O₂/H/O in different addition options.**

Here, "corresponding amount" implies that the molar ratio of added hydrogen to oxygen is 2 to 1. Adiabatic flame temperatures, burning velocities, and pollutant species concentrations of these mixtures have been calculated for all 9 mixtures at equivalence ratios from 0.6 to 1.3.

Increases in the burning velocity with the addition of molecular and atomic oxygen and hydrogen are forecasted because: 1) increased oxygen concentration increases the radical pool concentration by decreasing nitrogen dilution and increasing the temperature; 2) direct radical addition increases the reaction rate; 3) increased hydrogen concentration increases the thermal diffusivity [28]; 4) the higher burning velocity of $H_2$ directly affects burning velocity by increased thermal diffusion.

However, it can be seen in Figure 4 that the predicted increases in burning velocity over the base case (pure iso-octane/air) are not as high as expected. Even though hydrogen has a 5-fold higher burning velocity than iso-octane in stoichiometric mixtures, the increases in burning velocity with 5% and 10% $H_2$ addition are 1% and 5%, respectively. On the other hand, adding hydrogen leads to larger percentage increase in burning velocity away from $\phi = 1.0$. For example, a 15% increase has been predicted for 10% $H_2$ + iso-octane mixture at $\phi = 0.6$.

Similarly, $2H_2+O_2$ addition has been originally proposed to further improve the burning velocity by combining the effect of the higher burning velocity of $H_2$ and the effects of the enriched oxygen concentration.

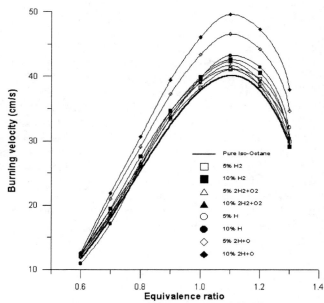

**Fig. 4. Comparison of burning velocities of different mixtures as a function of equivalence ratio ($T_0 = 298$, K $P_0 = 1$ atm).**

Similar to the $H_2$ addition case, magnitude of the resulting burning velocity enhancement is significantly less than expected. In contrast to the pure $H_2$ case, the $2H_2+O_2$ addition is relatively less sensitive to equivalence ratio; the increase in $S_u$ is 2-3% for the 5% $2H_2+O_2$ case and 3-5% for the 10% $2H_2+O_2$ case.

The expected increase in overall reaction rate with the addition of atomic hydrogen is confirmed. Surprisingly, for stoichiometric mixtures the increases are not much different than molecular hydrogen addition in both the 5% and 10% addition cases. On the other hand, the predicted increases for lean and rich mixtures are nearly two times higher than the values for $H_2$ addition.

In terms of burning velocity improvements, the most dramatic results are obtained by the addition of molecular hydrogen and oxygen simultaneously. Increases are 15% for 5% addition and 25% for 10% addition in fuel rich mixtures.

The adiabatic flame temperature of stoichiometric iso-octane/air flames is predicted to be 2282 K. For equivalence ratio between 0.6 and 1.2, all additives increase the adiabatic flame temperature, with the exception of 10% H at $\phi < 0.8$, as shown in Figure 5.

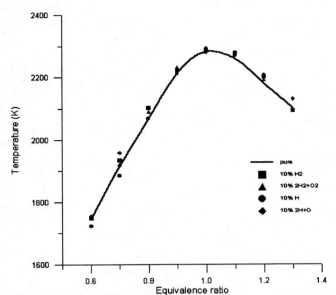

Fig. 5. Effects of different additives on the adiabatic flame temperature of iso-octane/air mixtures ($T_0$ = 298 K, $P_0$ = 1 atm).

Also, except 2H+O case, all other additives decreased the adiabatic flame temperature at $\phi$ = 1.3. At stoichiometric conditions with 10% addition, the largest temperature increase is predicted to be 12 K with 2H+O addition, followed by a 10 K increase with $2H_2+O_2$ addition, a 9 K increase with H addition, and 7 K increase with $H_2$ addition. It is also noted that the most effective additive on adiabatic flame temperature increase at the 5% level is $2H_2+O_2$, while 2H+O is the most effective for 10% additives.

Changes in predicted flame temperature profiles are depicted in Figure 6.

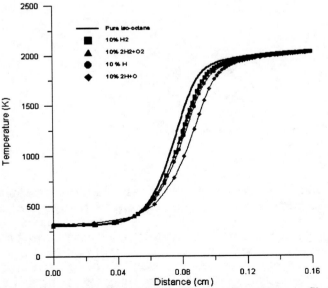

Fig. 6. Effects of different additives on the temperature profile of stoichiometric iso-octane/air mixtures.

For all additives, increasing the addition percentage caused an earlier start of the temperature rise in the preheat zone, but later in the reaction zone these

temperature profiles fell below the pure iso-octane curve. Finally, higher reaction rates in the recombination zone resulted in higher flame temperatures. The shapes of the temperature profiles indicate that hydrogen/oxygen additives have a small effect on the induction time, but the increase in reaction rates resulting in higher flame temperatures are responsible for the higher burning velocities with additives. The biggest change in temperature profile is caused by 2H+O addition.

The effects of different additives on the carbon monoxide concentration are plotted on Figure 7 as a function of the equivalence ratio.

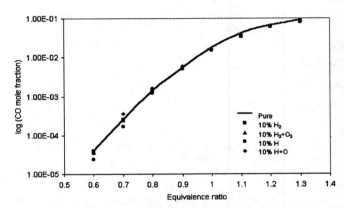

Fig. 7. Comparison of calculated CO emissions of different mixtures.

The quantitative effects of additives on CO emissions do not change with the percentage of addition. For this and clarity reasons, only the results for additives at the 10% level are shown. All additives result in decreased CO concentrations except mixtures of iso-octane with 2H+O at $0.6 < \phi < 1.0$. Increasing the amount of additives always result in higher reductions in CO emissions. Among all additive options, 10% H addition at $\phi$ = 0.6 provides the biggest reduction (20%) in CO emissions, followed by 2H+O (15% reduction) and $2H_2+O_2$ ($\phi$ = 0.6) additions. Amongst the rich mixtures, $H_2$ addition causes the highest CO reductions (16% reduction with 10% $H_2$ addition at $\phi$ = 1.2). Reductions in CO concentrations may be explained by decreased carbon concentration in the fuel. However, in some cases the amount of reduction is larger than this, indicating the importance of some kinetic interactions.

It is known that CO concentration reaches its maximum concentration at the flame front and is then oxidized within and after the reaction zone by the reaction:

$$CO + OH = CO_2 + H \qquad (2)$$

Therefore, CO concentration is mostly effected by the OH concentration in the post flame region. Every change in chemistry resulting in increase in OH concentration may enhance CO oxidation. To explain the possible interaction between CO and OH chemistry, normalized sensitivity coefficients in terms of mole fractions are computed. Figures 8 and 9 show the reactions to which CO mole fraction is most sensitive (at the flame and post-flame region in Figures 8 and 9, respectively) as a function of flame coordinate for a stoichiometric pure iso-octane flame. In Figure 8, the CO mole fraction starts to increase as soon as there is a change in temperature and reaches its maximum concentration close to the point of maximum temperature slope.

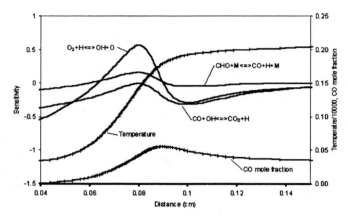

**Fig. 8. Normalized sensitivity coefficients of different reactions on CO concentration at the flame front as a function of flame coordinate ($\phi$ = 1.0, iso-octane/air).**

Carbon monoxide then steadily decreases to the end of the computational domain as shown in Figure 9.

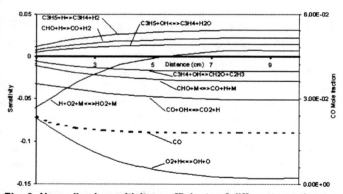

**Fig. 9. Normalized sensitivity coefficients of different reactions on CO concentration at the selected post-flame region of stoichiometric iso-octane/air flames**

Within the flame front, CO is generated as a result of a series of different fuel oxidation reactions. Among these reactions, the most influential one on CO generation is:

$$CHO + M = CO + H + M \qquad (3)$$

However, in the flame and post-flame regions some OH generating reactions are found to be as influential as direct CO involving reactions on CO concentrations. These effects are mostly due to the oxidation of CO by reaction (2), which is found to be the most influential reaction on CO mole fraction in the post-flame region. In

Figure 8, a negative sensitivity coefficient indicates that the reaction decreases the CO concentration. To strengthen these results, differences in OH concentrations are also investigated in this study.

The hydroxide radical mole fractions at the end of computational domain ($x$ = 10 cm) are shown in Figure 10.

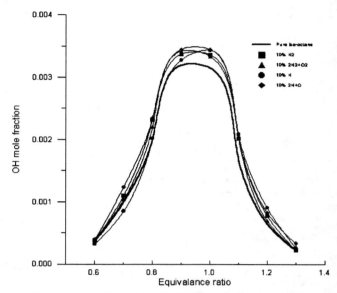

**Fig. 10. Comparison of simulated OH concentrations of different mixtures.**

Increases in OH concentrations are calculated for all additives at all equivalence ratios, except 10% H addition at $\phi < 0.7$. Increases in OH concentration with hydrogen/oxygen can be explained by the following mechanism:

$$O_2 + H = OH + O \qquad (4)$$

$$H_2 + O = OH + H \qquad (5)$$

$$2H + M = H_2 + M \qquad (6)$$

$$2O + M = O_2 + M \qquad (7)$$

$$O + H + M = OH + M \qquad (8)$$

$$O + H_2 = H + OH \qquad (9)$$

Since the high OH points correspond to the maximum CO reduction points, this appears to indicate that the increase in CO oxidation is due to the increase in OH mole fraction. For the range of conditions considered, increasing the amount of additive always lead to higher OH concentrations. In all cases, the percentage of OH mole fraction change is found to increase away from stoichiometry. The maximum differences in OH mole fractions from the base case occurred for the 2H+O addition case; that is, 40% at $\phi$ = 1.3 with 10% addition, and 30% at $\phi$ = 1.3 with 5% addition.

The effects of hydrogen/oxygen additives on NO concentrations are shown in Figure 11.

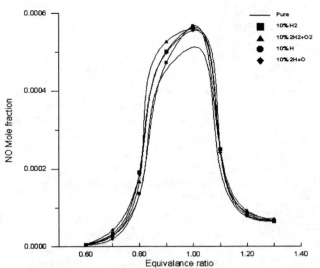

Fig. 11. Effects of hydrogen/oxygen additives on NO emissions.

High temperatures and high oxygen concentrations are believed to be the major factors affecting the NO formation rates [30]. For all additive cases, higher adiabatic flame temperatures are responsible for the increased NO concentrations. The addition of oxygen, on the other hand, results in both increased oxygen concentrations and increased adiabatic flame temperatures, leading to further increase in NO concentrations. This is particularly obvious for the $2H_2+O_2$ case near stoichiometry.

Sensitivity analysis has been performed to explain the effects of different reactions on NO mole fractions. The reactions influencing NO mole fraction are different in the flame region from those in the post-flame region, as shown in Figures 12 and 13, respectively.

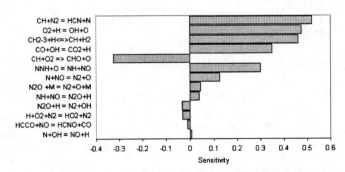

Fig. 12. Normalized sensitivity coefficients of different reactions on NO concentration at the flame front ($x$ = 0.09687 cm, $T$ = 1896 K) of stoichiometric iso-octane/air flames.

In the flame region, the first 6 reactions in Figure 12 are found to be most effective. However, their effects on total NO emissions are limit; only approximately 10% of the total NO is produced in the flame front. Post flame NO mechanisms are found to be more effective instead;

roughly 90% is produced in the post flame region. In post-flame conditions, the single most influential reaction on NO mole fraction is:

$$N + NO = N_2 + O \qquad (10)$$

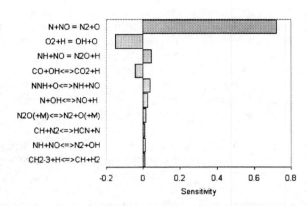

Fig. 13. Normalized sensitivity coefficients of different reactions on NO emissions at the selected post-flame region ($x$ =10 cm $T$ = $T_{ad}$ = 2295 K) of stoichiometric iso-octane/air flame.

Although $NO_x$ emissions are the sum of NO and $NO_2$, the final $NO_2$ concentrations for all mixtures and all equivalence ratios are predicted to be less than 1% of the final ($x$ = 10 cm) NO concentrations as illustrated in Figure 14.

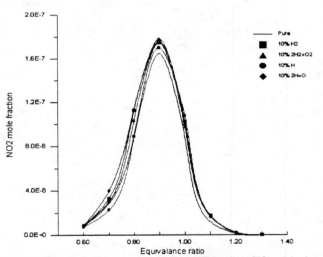

Fig. 14. Effects of hydrogen/oxygen additives on $NO_2$ emissions.

Also, the effects of additives showed similar trends between $NO_2$ and NO concentrations. The additives examined in this work always increased $NO_2$ levels.

## CONCLUSION

A detailed chemical kinetics simulation is utilized to produce accurate predictions on the effects of hydrogen and oxygen additives on freely propagating, laminar, premixed iso-octane/air flames. Adiabatic flame temperatures, burning velocity, and CO, OH, and NOx emissions are calculated for different types of additives and different equivalence ratios.

All types of hydrogen/oxygen additives considered are beneficial in terms of burning velocity enhancements. All additives lead to increased adiabatic flame temperatures, and increase in adiabatic flame temperature increases with the quantity added. It is found that increased OH concentrations might be the reason behind the reduction in CO concentrations. Up to 40% reduction in CO is calculated in lean mixtures of iso-octane/air with the addition of 10% atomic hydrogen. Increases in adiabatic flame temperatures and oxygen concentrations ($2H_2+O_2$ case) are found to be most effective in increasing NO and $NO_2$ emissions. However, increases in $NO_x$ emissions are predicted to be smaller if these additives are employed in lean mixtures.

Additional research is needed in order to better understand and explain the kinetic mechanisms of iso-octane/air combustion, especially near the lean and rich flammability limits. It should also be noted that low temperature kinetics is probably a weak point of recent higher hydrocarbon mechanisms because of the uncertainties in reaction rate parameters and omission of some reaction pathways.

## ACKNOWLEDGMENTS

This project is funded by Environmental Science and Technology Alliance Canada (ESTAC). The authors are particularly grateful to Mr. A. M. McDowell of ESTAC, Dr. A. Buckley of Ontario Centre for Environmental Technology Advancement, and Mr. S. Legedza of Enbridge Consumer Gas.

## REFERENCES

1. Curran, H. J., Gaffuri, P., Pitz W. J., and Westbrook, C. K., Combust. Flame 114:149 (1998).
2. Roberts, C. E., Matthews, R. D., and Leppard, W. R. (1996). SAE Technical Paper 962107.
3. Hasse, C., Bollig, M., Peters, N., and Dwyer, H. A., Combust. Flame 122:117 (2000).
4. Ranzi, E., Dente, M., Goldaniga, A. Bozzano, G., and Faravelli, T., Prog. Energy. Combust. Sci. 27:99 (2001).
5. Glaude, P. A., Battin-LeClerc, F., Fournet, R., Warth, V., Come, G. M., and Scacchi G., Combust. Flame 122:451 (2000).
6. Griffiths, J. F., Prog. Energy. Combust. Sci. 21:25 (1995).
7. Nehse, M., Warnatz, J., and Chevalier, C., XXVI Symp.(Int.)on Comb., 1996, p. 773.
8. Jones, H. R. N., and Leng, J., Combust. Flame 99:404 (1994).
9. Raman, V., Hansel, J., Kielian, D., Lynch, F., Ragazzi, R., and Wilson, B., 26th International Symposium on Automotive Technology and Automation, Aachen, Germany 1993.
10. Karim, G. A., Wierzba, I., and Al-Alousi, Y., Int. J. Hydrogen Energy 21:625 (1996).
11. Barbe, P., Martin, R., Perrin, D., and Scacchi, G., Int. J. Chem. Kinet. 28:829 (1996).
12. Gülder, Ö. L., Snelling, D. R., and Sawchuk, R. A., XXVI Symp. (Int.) on Comb., 1996, p. 2351.
13. Uykur, C., Henshaw, P. F., Ting, D. S.-K., and Barron, R. M., Int. J. Hydrogen Energy 26:265 (2001).
14. Nasser, S. H., Morris, S., and James, S. (1998). SAE Technical Paper 982561.
15. Flurchick, K., and Etters, R. D., AIAA Journal 23:981 (1985).
16. Choudhuri, A. R., and Gollahalli, S. R., Int. J. Hydrogen Energy 25:451 (2000).
17. Kee, R. J. et al., 2000, "CHEMKIN Collection, Release 3.6," Reaction Design, Inc., San Diego, CA.
18. Peters, N., Abschlußbericht zum DFG Forschungsvorhaben Pe 241/9-2, available on the World Wide Web http://www.itm.rwth-aachen.de, RWTH-Aachen, Institut für Technische Mechanik, D-52056 Aachen, Germany.
19. Smith, G. P., Golden, D. M., Frenklach, M., Moriarty, N. W., Eiteneer, B., Goldenberg, M., Bowman, C. T., Hanson, R. K., Song, S., Gardiner, Jr. W. C., Lissianski, V. V., and Qin, Z., http://www.me.berkeley.edu/gri_mech/.
20. Ern, A., and Giovangigli, V., J. Comput. Phys. 120:105 (1995).
21. Davis, S. G., and Law, C. K., Combust. Sci. Technol. 140:427 (1998).
22. Gülder, Ö. L., XIX Symp. (Int.) on Comb., 1982, p. 275.
23. Gibbs, G. J., and Calcote, H. F., J. Chem. Eng. Data 4:226 (1959).
24. Bradley, D., Hicks, R. A., Lawes, M., Sheppard, C. G. W., and Wooley, R., Combust. Flame 115:126 (1998).
25. Van Maaren, A, Thung, D. S., and De Goey, L. P. H., Combust. Sci. Technol. 96:327 (1994).
26. Zhu, D. L., Egolfopoulos, F. N., and Law, C. K., XXII Symp.(Int.) on Comb., 1988, p. 1537.
27. Muller, U. C., Bolig, M., and Peters, N., Combust. Flame 108:349 (1997).
28. Glassman, I. Combustion, Academic Press, San Diego, 1996, p. 125.
29. Curran, H. J., Pitz, W. J., Westbrook, C. K., Callahan, C. V., and Dryer, F. L., XXVII Symp. (Int.) on Comb., 1998, p. 379.
30. Heywood, J. B. Internal Combustion Engine Fundamentals. McGraw-Hill, New York, 1988, p. 572.

## CONTACT

Dr. Andrzej Sobiesiak: University of Windsor, Mechanical, Automotive and Materials Engineering, Windsor, Ontario, Canada N9B 3P4, Tel: (519) 253 3000 ext 3886, Fax: (519) 973 7007, e-mail: asobies@uwindsor.ca.

## NOMENCLATURE

$P_0$: Initial Pressure [atm]
$P_d$: Datum pressure [atm]
$S_u$: Laminar burning velocity [cm/s]
$S_{u,d}$: Datum laminar burning velocity [cm/s]
$T_{ad}$: Adiabatic flame temperature [K]
$T_0$: Initial Temperature [K]
$T_d$: Datum temperature [K]
$x$: Distance [cm]
$\alpha, \beta$: Constants
$\phi$: Fuel/Air Equivalence Ratio.

2002-01-1711

# Atomization Characteristics for Various Ambient Pressure of Dimethyl Ether (DME)

**Mitsuharu Oguma and Gisoo Hyun**
New Energy and Industrial Technology Development Organization, NEDO

**Shinichi Goto**
National Institute of Advanced Industrial Science and Technology, AIST

**Mitsuru Konno and Shuichi Kajitani**
Ibaraki University

## ABSTRACT

Recently, dimethyl ether (DME) has been attracting much attention as a clean alternative fuel, since the thermal efficiency of DME powered diesel engine is comparable to diesel fuel operation and soot free combustion can be achieved. In this experiment, the effect of ambient pressure on DME spray was investigated with observation of droplet size such as Sauter mean diameter (SMD) by the shadowgraph and image processing method. The higher ambient pressure obstructs the growth of DME spray, therefore faster breakup was occurred, and liquid column was thicker with increasing the ambient pressure. Then engine performances and exhaust emissions characteristics of DME diesel engine were investigated with various compression ratios. The minimum compression ratio for the easy start and stable operation was obtained at compression ratio of about 12. The brake thermal efficiencies with various compression ratio from 12 to 17.7 (original compression ratio) were almost same level when compared to the original compression ratio of the engine. When the engine fueled with DME is operated at low compression ratio, the theoretical thermal efficiency decrease. However, the high combustion efficiency, the increased degree of constant volume and the lower heat loss to cooling water make up for the decreased theoretical thermal efficiency. Furthermore, the low THC and CO emissions and engine noise were comparable to original compression ratio, and the reduction of NOx emission was achieved.

## INTRODUCTION

Global and urban environmental problems are caused by the rapid increase of carbon dioxide ($CO_2$) and harmful exhaust emissions due to the burning of fossil fuels. Available oil deposits are calculated to last only about 42 years, so we must convert from petroleum fuel to an environmentally safe and renewable new fuel.

Recently, dimethyl ether (DME) has been attracting much attention as a clean alternative fuel[1-6]. Table 1 shows characteristics of DME compared to the propane and diesel fuel. The vapor pressure of DME is comparable to propane and DME liquefy easily with a few pressurization. However, its lower heating value is about 60 % of diesel fuel and viscosity is very low.

DME has a high cetane number enough to operate diesel engine, the thermal efficiency of DME powered diesel engine is comparable to diesel fuel operation and soot free

Table 1 Characteristics of DME, propane and diesel fuel

| Property | DME | Propane | Diesel fuel |
|---|---|---|---|
| Chemical formula | $(CH_3)_2O$ | $C_3H_8$ | $C_nH1.8_n$ |
| Boiling point [K] | 248.1 | 231 | 453-643 |
| Cetan number | >>55 | - | 57.8 |
| Lower heating value [kJ/kg] | 28430 | 46360 | 43200 |
| Vapor pressure [MPa] at 293[K] | 0.53 | 0.83 | <0.001 |
| Liquid viscosity [cSt] | $0.18(\times10^{-3})$ | | $2.4\text{-}4.6(\times10^{-3})$ |

combustion can be achieved. However, the DME diesel engine has some problems such as the NOx emissions and low maximum power output and so on.

Atomization characteristics and mixture formation processes of DME might play an important role on engine performances and exhaust emissions. Particularly, it seems that atomization of DME has a great influence on ambient pressure because of its low boiling point and vapor pressure.

Therefore in this experiment, the effect of ambient pressure on DME spray was investigated with image observation and measurement of the droplet size by the shadowgraph and image processing method. Then engine performances and exhaust emissions characteristics of DME diesel engine were investigated with various compression ratios.

## EXPERIMENTAL APPARATUS AND PROCEDURES

### 2.1 Experimental apparatus

Figure 1 shows schematic of spray experimental apparatus used in this study. The spray experimental system (Greenfield system) based on shadowgraph was used for taking spray images, and it was made up of

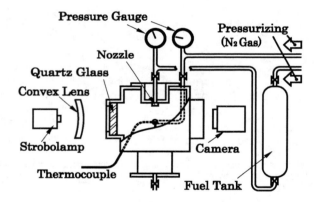

Fig.1 Schematic diagram of the spray experiment

strobe lamp, CCD camera, constant volume chamber and frame-grabber. Major specifications of Greenfield system and experimental conditions are listed in Table 2.

Figure 2 presents a schematic diagram of the engine experiment. The test engine used in this study is a four-stroke, single cylinder, naturally aspirated, direct injection diesel engine (Yanmar Diesel Corporation NFD-13K).

DME is gas state at the normal atmospheric pressure and temperature, so the fuel lines were modified to the pressure-resistant type and DME was pressurized by nitrogen gas to keep its liquefied state. The specifications of the engine and experimental conditions

are showed in Table 3. The compression ratios of the test engine were changed by using the gaskets with different thickness.

Table 2 Experimental conditions and specifications

| Measurable droplet range [μm] | 5-6000 |
|---|---|
| Strobo flash duration [μs] | ≤0.5 |
| CCD image resolution [pixels (mm)] | 640 x 480 (7.1 x 5.3) |
| Injector type | Single hole nozzle |
| Nozzle diameter [mm] | 0.3 |
| Length/diameter [mm] | 5.0 |
| Fuel type | DME |
| Fuel injection pressure [MPa] | 5.0 |
| Ambient gas pressure [MPa] | 0.5, 1.0, 1.5, 2.0 |
| Ambient gas temperature [K] | 298 |

### 2.2 Experimental procedures

In experiment of DME spray, liquid DME was injected from single hole (φ0.3 mm) nozzle into the constant volume chamber, which was also pressurized by air or nitrogen to change the ambient pressure. In this experiment, ambient pressures were set at 1.0 and 1.5

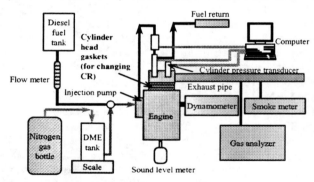

Fig.2 Schematic diagram of the engine experiment

Table 3 Engine parameters and experimental conditions

| Bore x Stroke [mm] | 92 x 96 |
|---|---|
| Rated output [kW/rpm] | 8.09/2400 |
| Injection pump plunger [mm dia.] | 8 (Bosch type) |
| Injector [mm hole dia] | 0.26 x 4 holes |
| Original static injection timing [°CA ATDC] | -17.0 |
| Nozzle opening pressure [MPa] | 19.6 |
| Engine speed [rpm] | 960 |
| Compression ratio | 17.7(Original), 16.87, 16.12, 13.89, 13.08, 12.36 |

MPa to equalize the density of in-cylinder working medium of test engine at TDC with compression ratio of 17.7 and 12 respectively. The ambient pressures were also set at 0.5 and 2.0 MPa to compare the results. The droplet equivalent diameter of SMD (Sauter mean diameter) was estimated based on equivalent circular area method [7]. Successive 60 frames of images were taken to obtain droplet size at each measurement condition.

In the case of engine experiment, the engine was operated at a constant speed of 960 rpm. The cylinder pressures and the brake specific fuel consumptions were measured for various compression ratios to investigate the effect of compression ratio on combustion characteristics of DME diesel engine. Also, the exhaust gas analyzer (HORIBA MEXA-8200) was used to measure NOx, total hydrocarbon (THC), CO and $CO_2$.

## RESULTS AND DISCUSSION

EFFECT OF AMBIENT PRESSURE ON DME SPRAY

Figure 3 shows saturated vapor line of DME. In this study, DME injected with injection pressure at 5.0 MPa

into an ambient pressures from 0.5 to 2.0 MPa, and ambient

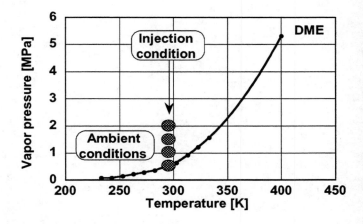

Fig.3 Vapor pressure curve of DME

temperature at 298 K. Except for the ambient pressure at 0.5 MPa, it is clear that the DME does not cross the saturated vapor line and so it seems that the DME is liquid at these ambient conditions.

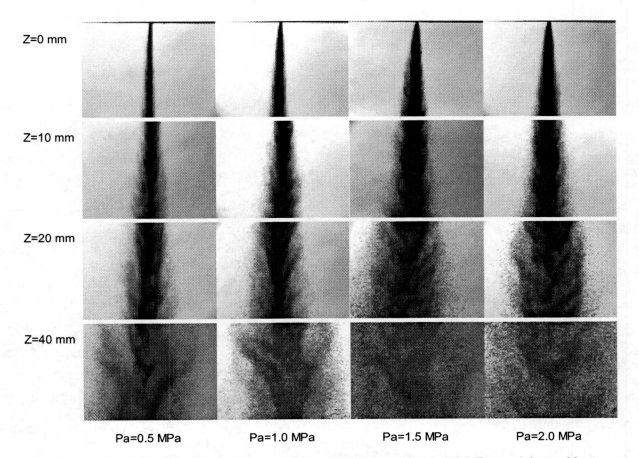

Fig.4 Image of DME spray when the fuel injection pressure of $P_{inj}$ is 5.0 MPa, and the ambient pressures of $P_a$ are 1.0 and 1.5 MPa

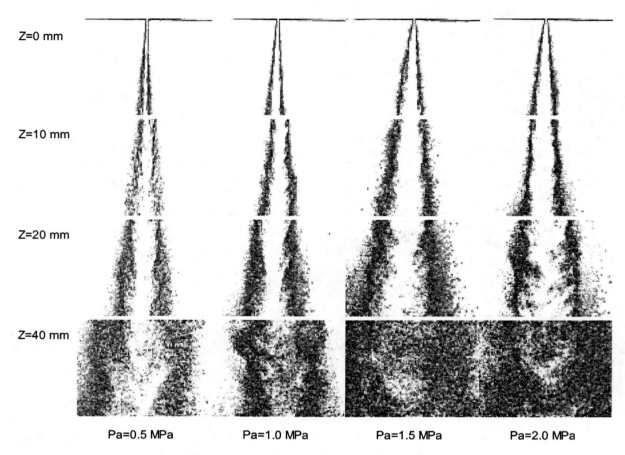

Z=0 mm

Z=10 mm

Z=20 mm

Z=40 mm

Pa=0.5 MPa     Pa=1.0 MPa     Pa=1.5 MPa     Pa=2.0 MPa

Fig.5 Modified spray images using computer software to sharpen the edge of the DME spray

Figure 4 shows some spray images of DME when the fuel injection pressure of $P_{inj}$ is 5.0 MPa, and the ambient pressures of $P_a$ are 0.5, 1.0, 1.5 and 2.0 MPa, respectively. In cases of ambient pressures higher than 1.0 MPa, a lot of small fuel droplets were to be seen around ligament. On the other hand, there is misty area in DME spray of $P_a$=0.5 MPa. This area seems gaseous DME. The thickness of liquid column increased with raising the ambient pressure, and liquid column of low ambient pressure remained until axial distance from nozzle tip as Z=40 mm.

Figure 5 was modified from Fig.4 by using computer software to sharpen the edge of the DME spray. The liquid column was thicker with increasing the ambient pressure. It seems that the higher ambient pressure obstructs the growth of DME spray, therefore faster breakup was occurred, and liquid column was thicker with increasing the ambient pressure.

Figure 6 shows effect of ambient pressure on Sauter mean diameter (SMD) of the DME spray with ambient pressure as 0.5, 1.0, 1.5 and 2.0 MPa. All of them, SMD

were observed from 150 to 170 μm at nozzle tip, then SMD decreased suddenly with increasing axial distance

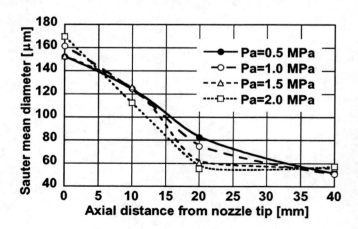

Fig.6 Effect of ambient pressure on Sauter mean diameter (SMD)

from nozzle tip (Z) until Z=20 mm. In case of Z=20 mm, SMD of DME spray decreased with raising the ambient pressures. After Z=20 mm, the decrease ratio was changed as a gentle slope.

In summary, the effect of ambient pressure on DME spray is comparable to liquid fuels when the ambient temperature is not high enough to vaporize the liquid DME.

It seems that the effect of ambient temperature on DME spray evaporation is also an important parameter. Therefore, the effect of high ambient temperature should be investigated.

## EFFECT OF COMPRESSION RATIO ON DME DIESEL ENGINE PERFORMANCES AND EXHAUST CHARACTERISTICS

Figure 7(a) shows pressure-time histories of the DME diesel engine for various compression ratios. The rates of heat release were calculated using these pressure data, and are plotted in Fig.7(b). The engine was able to be started and operated with a compression ratio as low as about 12 without any problems. The peaks of pre-mixed combustion increased with lowering the compression ratio for both of the fuels, but DME was relatively lower than diesel fuel because of its ignition lag were shorter than diesel fuel as shown in Fig.8(a). It seems that the lower boiling point and vapor pressure of DME effect the

decreasing of ignition lag. Therefore, maximum of $(dP/d\theta)$, shown at Fig.8(b), operated with DME was lower than that of diesel fuel at low compression ratio. Figure 8(c) shows the effect of compression ratio on sound pressure level of the DME diesel engine compared to diesel fuel operation. The sound pressure levels of DME operation were low even at low compression ratios, because of low pre-mixed combustion and maximum of $(dP/d\theta)$ with DME compared to the diesel fuel operation.

Figure 9 show the effect of compression ratio on the (a) CO and (b) THC emissions of the DME diesel engine compared to the diesel fuel operation. The emission of CO and THC were very low when operated with DME and did not increase with lowering the compression ratio.

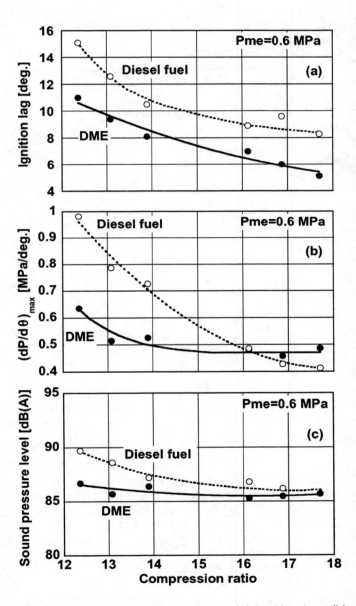

Fig.8 Effect of compression ratio on (a) ignition lag, (b) $(dP/d\theta)_{max}$ and (c) sound pressure level

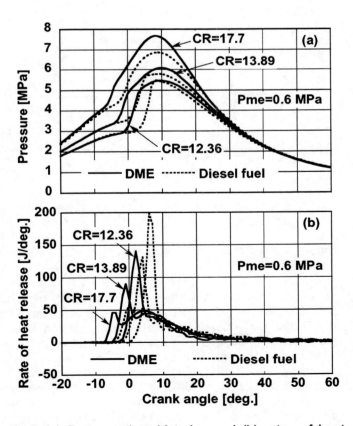

Fig.7 (a) Pressure-time histories and (b) rates of heat release for various compression ratios

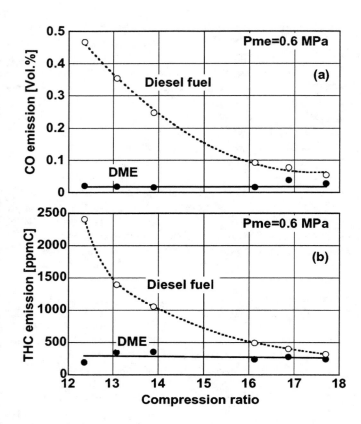

Fig.9 Effect of compression ratio on (a) CO and (b) THC emission

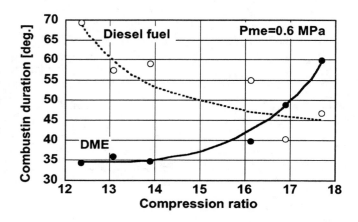

Fig.11 Effect of compression ratio on combustion duration

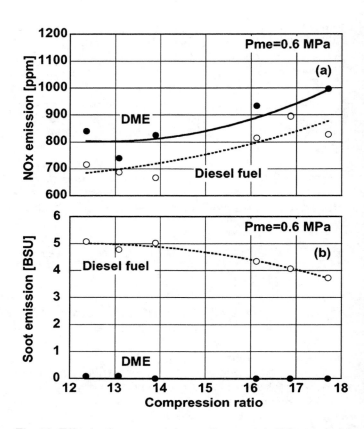

Fig.12 Effect of compression ratio on (a) NOx and (b) soot emission

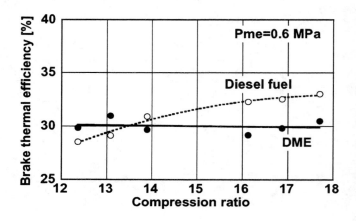

Fig.13 Effect of compression ratio on brake thermal efficiency

Figure 10 shows the combustion observation of DME diesel engine with compression ratio at 17.7 and 12.36 by shadowgraph method with their rates of heat release. In case of original compression ratio (17.7), the combustion started at −8 deg. ATDC. At this timing, DME spray was still growing and did not reach combustion chamber wall. On the other hand, operating with compression ratio at 12.36, combustion started after arrival of DME spray on chamber wall. Collision of DME spray with chamber wall made forming a good DME-air mixture, which may be due to the combustion duration of DME, as sown in Fig.11, was short at low compression

ratio. So that CO and THC emissions with DME were at low level.

Figure 12(a) shows the effect of compression ratio on NOx emissions of the DME diesel engine compared to the diesel fuel operation. At any compression ratios, higher engine out NOx emissions were observed with DME when compared to the diesel fuel operation. This is the most important problem of DME powered diesel engine. However, both DME and diesel fuel operations showed that the NOx emissions decreased with lowering compression ratio. Especially, at low compression ratio, NOx reduction became more effective for DME than diesel fuel.

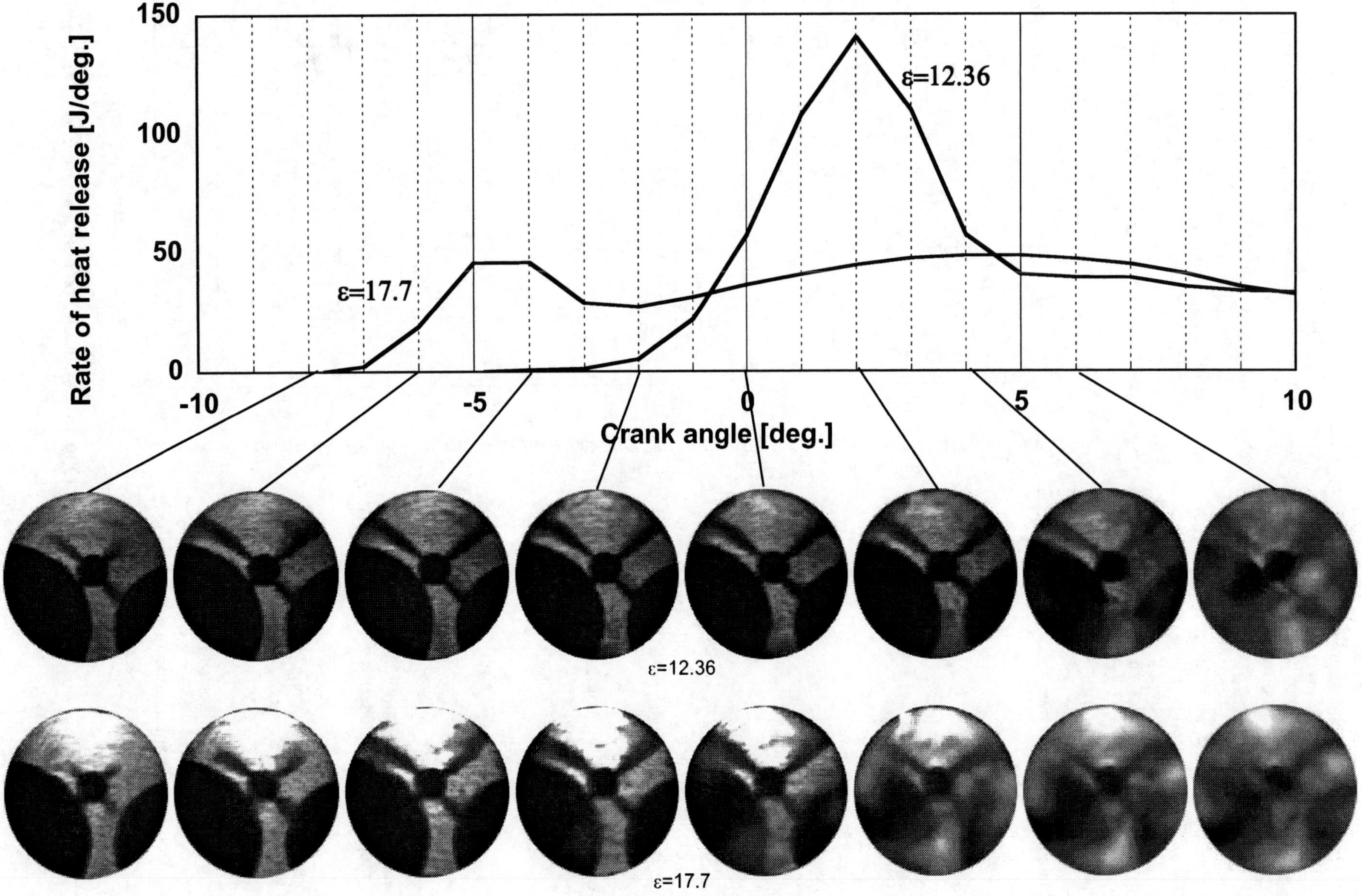

Fig.10 The combustion observation of DME diesel engine with compression ratio at 17.7 and 12.36 by shadowgraph method

Figure 12(b) shows the effect of compression ratio on soot emission. When the engine was operated with diesel fuel, the soot emissions increased with lowering the compression ratio. However, in case of DME, the soot emissions were negligible at any compression ratios.

Figure 13 shows the effect of compression ratio on brake thermal efficiency of DME diesel operation compared with diesel fuel operation. When the engine was operated with diesel fuel, the brake thermal efficiency decreased with lowering the compression ratio. However, in case of DME, the brake thermal efficiencies at any compression ratios were comparable to the original compression ratio (17.7).

Brake thermal efficiency can be shown roughly as follow.

$$\eta_e = \eta_{th}\eta_{glh}\eta_u(1-\phi_w)\eta_m \qquad (1)$$

Where
$\eta_e$ : Brake thermal efficiency
$\eta_{th}$ : Theoretical thermal efficiency of Otto cycle
　　($\eta_{th}=1-1/\varepsilon^{k-1}$)
$\eta_{glh}$ : Degree of constant volume
$\eta_u$ : Combustion efficiency
$\phi_w$ : Loss to cooling water
$\eta_m$ : Mechanical efficiency

Theoretical thermal efficiency decreases with lowering compression ratio, that is in case of DME operation, some efficiency made up the reduction of theoretical thermal efficiency.

Figure 14 shows the effect of compression ratio on mechanical efficiency. The mechanical efficiency was calculated by follow.

$$\eta_m = \frac{P_{me}}{P_{mi}} \qquad (2)$$

Where
$P_{me}$ : Brake mean effective pressure
$P_{mi}$ : Indicated mean effective pressure

Both of DME and diesel fuel operations, the mechanical efficiencies was the same level and increased as lowering the compression ratio.

Figure 15(a) shows the effect of compression ratio on combustion efficiency calculated by the concentrations of CO and THC emission in the exhaust gas using following equation.

$$\eta_u = 1 - \frac{CO+THC}{BSFC} \qquad (3)$$

Where
CO　: CO emission　　　　　　　　　[g/kWh]
THC　: THC emission　　　　　　　　[g/kWh]
BSFC: Brake specific fuel consumption [g/kWh]

The combustion efficiencies of diesel fuel operation decreased with lowering the compression ratio. However, these DME operations were at high level constantly. Better fuel/air mixture and low boiling point and vapor pressure of DME in low compression ratio seems to be the reason for high combustion efficiency.

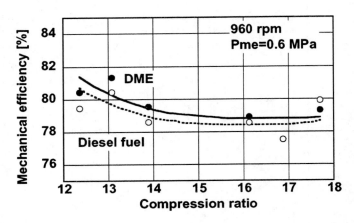

Fig.14　Effect of compression ratio on mechanical efficiency

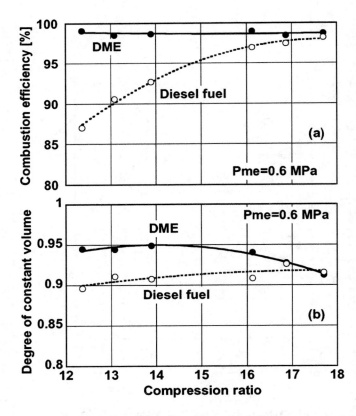

Fig.15　Effect of compression ratio on (a) combustion efficiency and (b) degree of constant volume

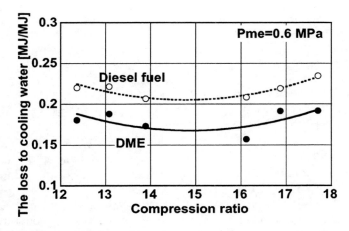

Fig.16 Effect of compression ratio on loss to cooling water

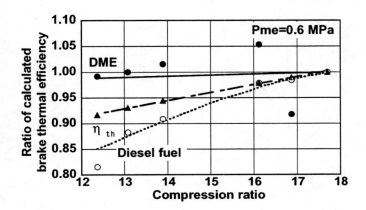

Fig.17 Effect of compression ratio on ratio of theoretical and calculated brake thermal efficiency

Figure 15(b) shows the effect of compression ratio on the degree of constant volume, and Fig.16 shows on the loss to cooling water. When the compression ratios are constant, the thermal efficiency reaches the peak in constant volume cycle. The degree of constant volume shows the decrease rates of thermal efficiency compared with that of constant volume cycle. When the engine was operated fueled with DME, the degree of constant volume increased with lowering the compression ratio, but that of diesel fuel decreased. The heat loss to cooling water operated with DME was lower than that with diesel fuel for all of the compression ratios. Because soot emission of DME operation were negligible at any compression ratios, that is heat loss of radiation from DME combustion seems lower than that of diesel fuel.

Figure 17 shows the relationship between the decrease rates of theoretical thermal efficiency, brake thermal efficiency calculated using the above efficiencies and compression ratio. The decrease rate was defined as dividing the results of the engine by the reference original compression ratio. The decrease rate of the brake thermal efficiency of diesel fuel was much greater than the decrease rate of the theoretical thermal efficiency. However, with DME it was negligible although the theoretical thermal efficiency decreased 8 % if the compression ratio decreased from 17.7 to about 12. In summary, when the engine fueled with DME was operated at low compression ratio, it seems that the high combustion efficiency, the increased degree of constant volume and the low heat loss to cooling water made up the decreased theoretical thermal efficiency.

## CONCLUSION

Spray behavior and atomization characteristics of DME have been investigated by using the shadowgraph and image processing method. Especially, the effect of an ambient pressure on DME spray has been examined. Moreover, an DI DME diesel engine has been operated with various compression ratios, and compared engine performances and exhaust emissions characteristics with diesel operation. The results obtained from this study can be summarized as follows:

1. DME is liquid state in the ambient conditions of this work. Therefore, liquid column is observed to the downstream in spite of evaporation characteristics of DME, and a lot of small droplets exist around ligament.
2. Higher ambient pressure leads to faster breakup, because of growing up the shear force owing to higher ambient resistance.
3. The effect of ambient pressure on DME spray characteristics such as spray angle and SMD are comparable to other liquid fuels, because the ambient temperature is not enough high to vaporize the DME fuel.
4. The minimum compression ratio obtained at around 12 is also suitable for easy start and stable operation of the DME fueled diesel engine. The brake thermal efficiency of the DME engine is almost same level for various compressions from 12.36 to 17.7.
5. At any compression ratios of the low-compression ratio DME engine, smokeless operation and low THC and CO emissions are achieved. NOx emissions decrease with lowering the compression ratio.
6. When the engine fueled with DME is operated at low compression ratio, the theoretical thermal efficiency decrease. However, the high combustion efficiency, the increased degree of constant volume and the lower heat loss to cooling water make up for the decreased theoretical thermal efficiency.

## ACRONYMS, ABBREVIATIONS

| | |
|---|---|
| **DME** | Dimethyl ether |
| **SMD** | Sauter mean diameter |
| **THC** | Total hydrocarbon |
| **CO** | Carbon oxide |
| **CO₂** | Carbon dioxide |
| **NOx** | Nitrogen oxide |
| **ATDC** | After piston top dead center |
| **BSFC** | Brake specific fuel consumption |

# REFERENCES

1. Sorenson, S.C. and Mikkelsen, S-E., "Performance and Emissions of a 0.273-Liter Direct-injection Diesel Engine Fueled with Neat Dimethyl Ether," SAE Paper 950064, 1995.
2. Kapus, P. and Ofner, H., "Development of Fuel-injection Equipment and Combustion System for DI Diesels Operated on Dimethyl Ether," SAE Paper 950062, 1995.
3. Fleisch, T., McCarthy, C., Basu, A., Udovich, C., Charbonneau, P., Slodowske, W., Mikkelsen S-E., and McCandless, J., "A New Clean Diesel Technology: Demonstration of ULEV Emissions on a Navistar Diesel Engine Fueled with Dimethyl Ether," SAE Paper 950061, 1995.
4. Kajitani, S., Chen, Z., Konno, M., and Rhee, K.T., "Engine Performance and Exhaust Characteristics of a Direct-injection Diesel Engine Operated with DME," SAE Paper 972973, 1997.
5. Alam, M., Fujita, O., Ito, K., Kajitani, S., and Mitsuharu, O., "Performance of NOx Catalyst in a DI Diesel Engine Operated with Neat Dimethyl Ether," SAE Paper 1999-01-3599.
6. Konno, M., Kajitani, S., and Suzuki, Y., "Unburned Emissions from a DI Diesel Engine Operated with Dimethyl Ether," Proc. of the 15th Int. Comb. Eng. Symp. 9935167, pp.69-74, 1999.
7. Lee, D., Goto, S., Shakal, J. and Hong, S., "Characteristics of an LPG Spray in a Constant Volume Chamber", 7th Symposium (ILASS-JAPAN) on Atomization, pp.182-192, 1998.